電路板基礎技術手札

林定皓　編著

全華圖書股份有限公司

國家圖書館出版品預行編目資料

電路板基礎技術手札 / 林定皓編著. -- 初版. --
新北市 : 全華圖書, 2018.06
面 ; 公分
ISBN 978-986-463-853-6(平裝)

1.印刷電路

448.62 107008674

電路板基礎技術手札

作者 / 林定皓

發行人 / 陳本源

執行編輯 / 李文菁

封面設計 / 曾霈宗

出版者 / 全華圖書股份有限公司

郵政帳號 / 0100836-1 號

印刷者 / 宏懋打字印刷股份有限公司

圖書編號 / 06371

初版一刷 / 2018 年 08 月

定價 / 新台幣 300 元

ISBN / 978-986-463-853-6(平裝)

全華圖書 / www.chwa.com.tw

全華網路書店 Open Tech / www.opentech.com.tw

若您對書籍內容、排版印刷有任何問題，歡迎來信指導 book@chwa.com.tw

臺北總公司(北區營業處)
地址：23671 新北市土城區忠義路 21 號
電話：(02) 2262-5666
傳真：(02) 6637-3695、6637-3696

南區營業處
地址：80769 高雄市三民區應安街 12 號
電話：(07) 381-1377
傳真：(07) 862-5562

中區營業處
地址：40256 臺中市南區樹義一巷 26 號
電話：(04) 2261-8485
傳真：(04) 3600-9806

編者序

現今編輯各版本電路板製造技術簡述小冊，陸續更新發行已超過萬本。本書秉持不贅述、輕便、易讀的初衷，讀者閱讀後可隨時翻閱不會有負擔。內容撰寫爲基礎技術介紹，要談技術細節則必須針對專題。如果是一般性理解目的，本書內容應該已經足夠。想瞭解進階內容，可參閱封面裡套書介紹之相關書籍。

本書內容簡要圖文對照，搭配簡單英文，可輔助與國外人士溝通與閱讀。爲秉持「口袋書」概念，並不做過度增補，以免入門閱讀困難。

相同議題必然有各方不同見解，出書之際除了對過去曾給予指正、建議的先進、同業好友致謝，也期許此小冊的更新，能對產業持續發展小有助益。

景碩科技 林定皓

2018 年春謹識於台北

編輯部序

「系統編輯」是我們的編輯方針，我們所提供給您的，絕不只是一本書，而是關於這門學問的所有知識，它們由淺入深，循序漸進。

本書以手札的方式呈現，分為六個章節，文字內容淺顯，搭配常用的專業詞彙，閱讀簡潔又省力，且以中英對照，讓沒有理工背景的人都能輕易理解其中重點，不論入門、回顧、說明、溝通，都是最佳的隨手書。本書適用於電路板相關從業人員使用。

同時，本書為電路板系列套書 (共 10 冊) 之一，為了使您能有系統且循序漸進研習相關方面的叢書，我們分為基礎、進階、輔助三大類，以減少您研習此門學問的摸索時間，並能對這門學問有完整的知識。若您在這方面有任何問題，歡迎來函聯繫，我們將竭誠為您服務。

目 錄

1

電子設備的結構

Electronic Products Structure

電子設備的結構

電子設備製造程序，從半導體裸晶開始，先構裝成顆粒，或直接裝載到板面 (COB - Chip On Board)。複雜的部件經過構裝，仍會安裝到載體或載板，之後組裝成介面卡或模組，再進一步與主機板完成組合成爲產品。

雖然不是所有電子設備都遵循相同模式，但大致程序與架構很相似。如：行動電話、錄影機、數位相機、個人電腦、穿戴裝置都類似，只是複雜度與細密度不同罷了。

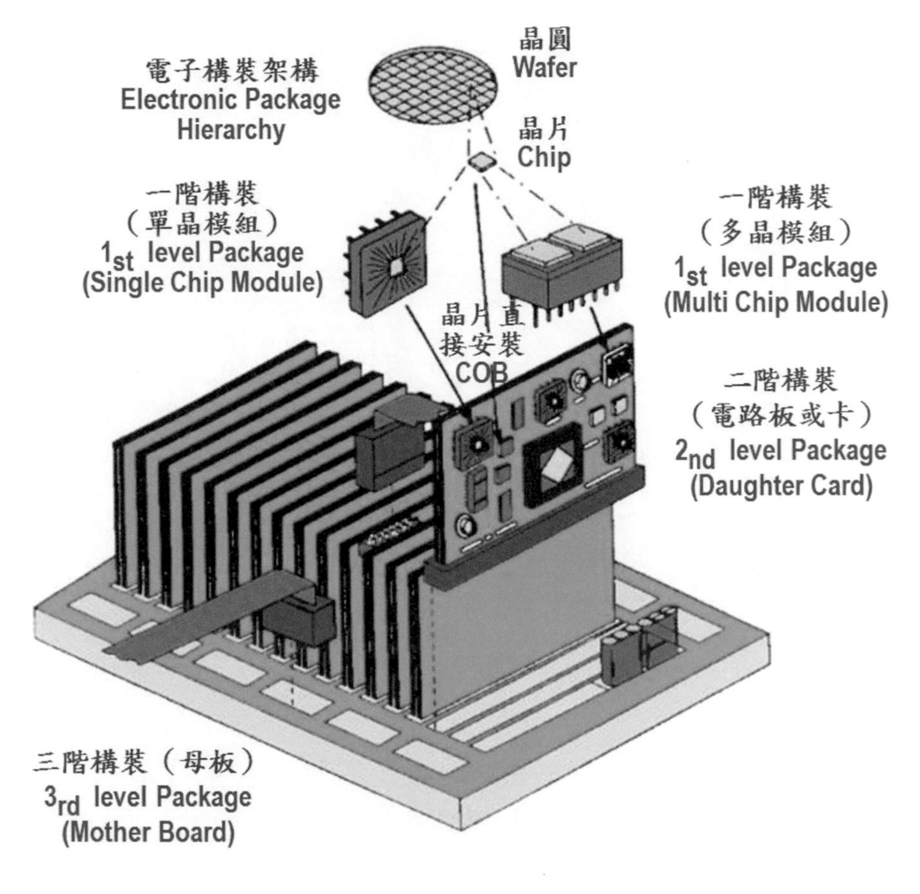

電子產品簡單的構裝階層 / Electronic devices MFG sequence

Electronic Products Structure

Electronic devices will make chip first; then attach them on carriers or substrate. Simpler ones might made by “Chip On Board”, the complicated ones still go through IC packaging & then assembled on cards or modules. And then, they will be installed or connected on main boards.

Different products will go unique way making, but the basic concept & procedure will be similar. Mobile Phone; VCR; DSC; PC and so on, basic architecture are similar, but the complexity and density are totally different.

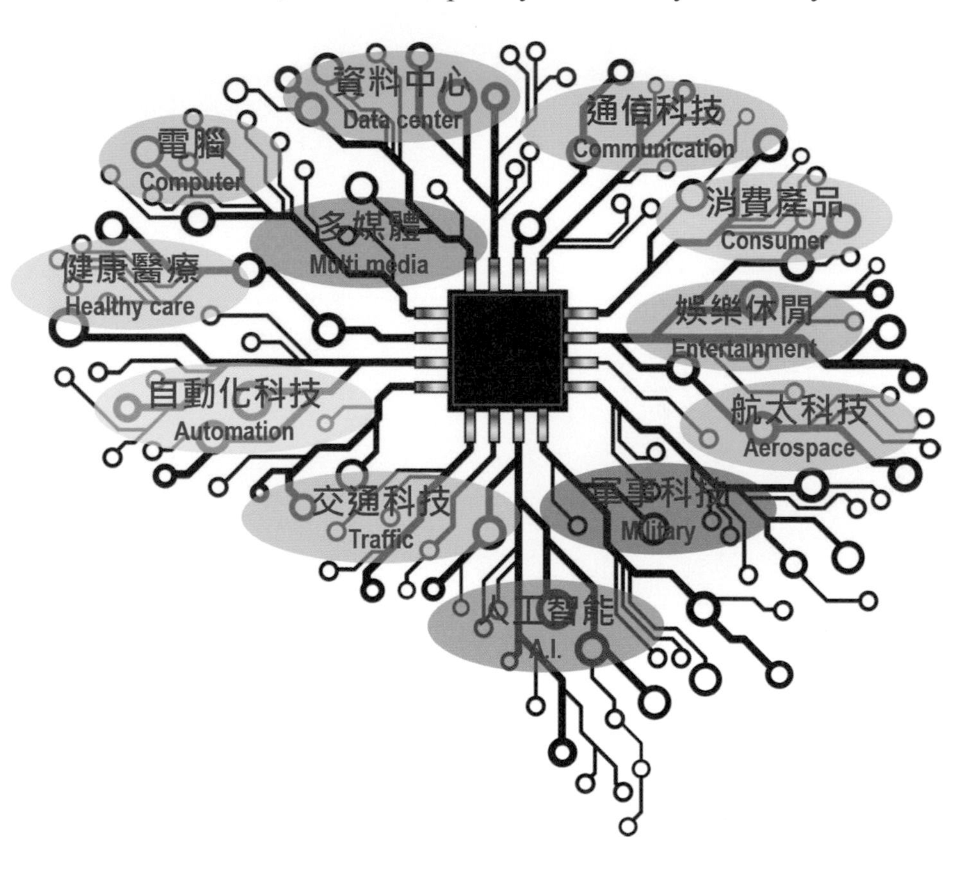

電路板讓產品有無限可能 / PCB make things happen.

2

電路板進化史

PCB Industry story

電路板進化史

1. PCB(Printed Circuit Board) 中文翻譯爲印刷電路板，也稱 PWB(Printed Wiring Board) 印刷線路板。該產品是以影像技術製作的電路產品。它取代了 1940 年代前，電器產品以銅配線製作的方式，使產品大量生產複製速度加快，產品體積縮小、品質穩定、方便生產、單價降低。
2. 早期電路板形式多樣化，曾將金屬融熔覆蓋在絕緣板表面，做出所要線路。1936 年後，製作方法轉向將覆蓋金屬絕緣基板，以耐蝕刻油墨作區域選別，將不要的區域以蝕刻去除，這種做法叫做“Subtractive Method”或“Print & Etch”。
3. 1960 年後，電唱機 / 錄音機 / 錄影機等，陸續採用“雙面貫通孔”電路板製造，於是耐熱及尺寸安定的“環氧樹脂基板”被大量使用，至今仍爲電路板製作主要樹脂基材。

銅線組裝對電路板組裝 / Copper wiring vs PCB assembly.

PCB Industry story

1. PCB is the abbreviation of "Printed Circuit Board", or known as PWB "Printed Wiring Board". They are made by photo lithography technology. It had replaced copper wiring & enables production & duplicating fast. It increased output & mobility, also make products stable & lower cost.
2. In early days, PCB has varieties feature & makers ever applied casting technology creating patterns on insulating plates. After 1936, they used resist ink to define patterns & then applied etching. This technology was so called "Subtractive Method" or "Print and etch" process.
3. In 60′s, Phonographs/Tape recorders/VCRs started using double sided PCBs with plated through-hole. As epoxy with stable thermal resistance & good dimension stubility, were used as the main resin material system for PCB, and kept still now.

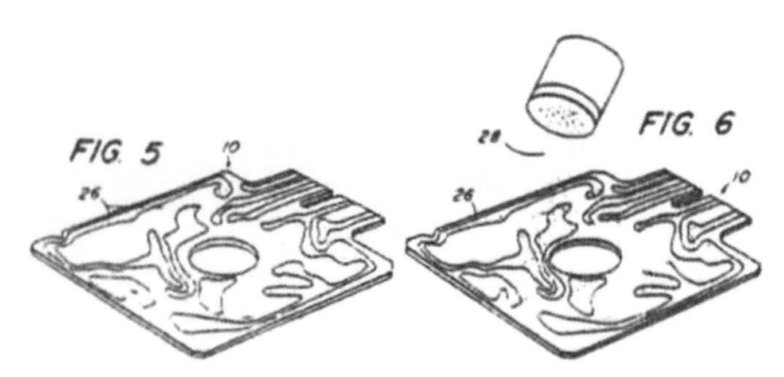

U.S. Patent 4,421,944. This figure shows the tacky ink been dusted with metal powder to enhance electrical conductivity.

美國專利 4,421,944 號顯示，以金屬粉粒灑在黏性油墨上增加導電度

一種早期電路板形式 / One old PCB style

電路板的分類

電路板分類，沒有絕對的方法與原則，可依據分類目的做不同考慮。分類原則，以能夠幫助了解與學習爲原則。

PCB Categories

PCB Category don't have firm rule & then can base on purpose doing that. The rule of thumb will depend on easier learning and benefit for aware.

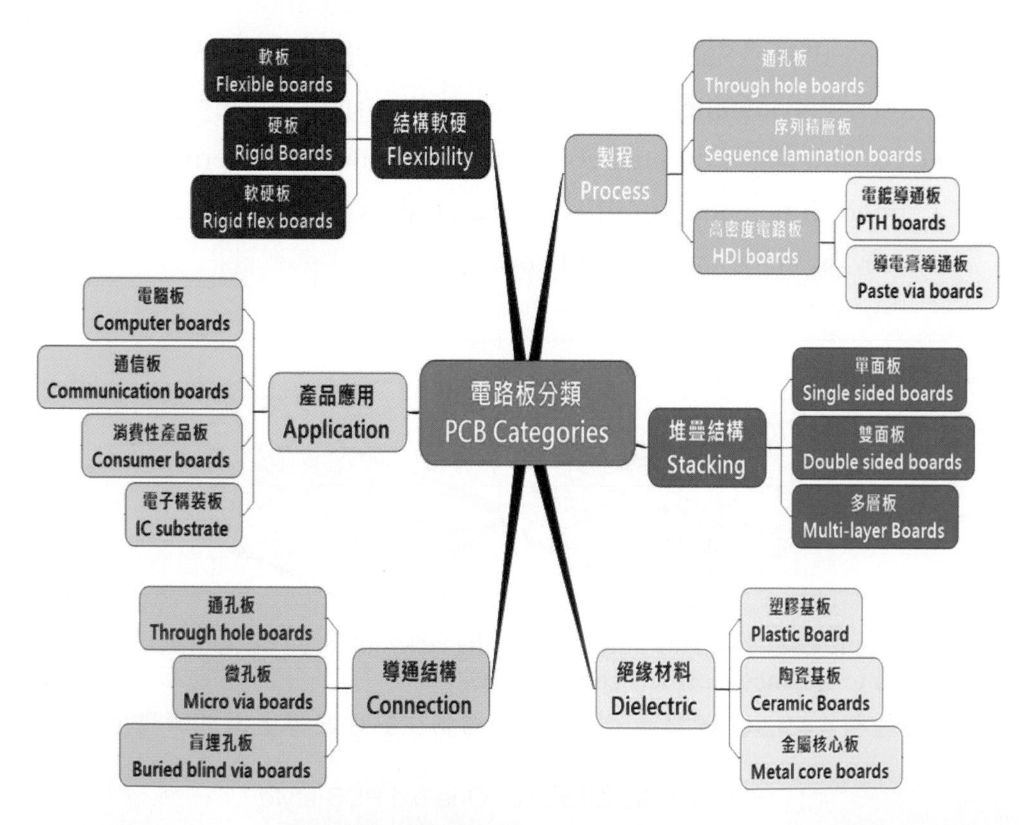

簡單的電路板分類 / Brief PCB category

3

電路板的應用

Applications of PCBs

電路板的應用

數位化工業急速發展，刺激了 PCB 量產化，由早期收音機、電視機、電唱機，代之而起的有：隨身聽、電算機、電腦、筆電、電子交換機、行動電話等接踵而至。

過去定義電路板的應用，習慣以 3C 描述類型，不過在高度整合發展下，已經很難再這樣區分了。

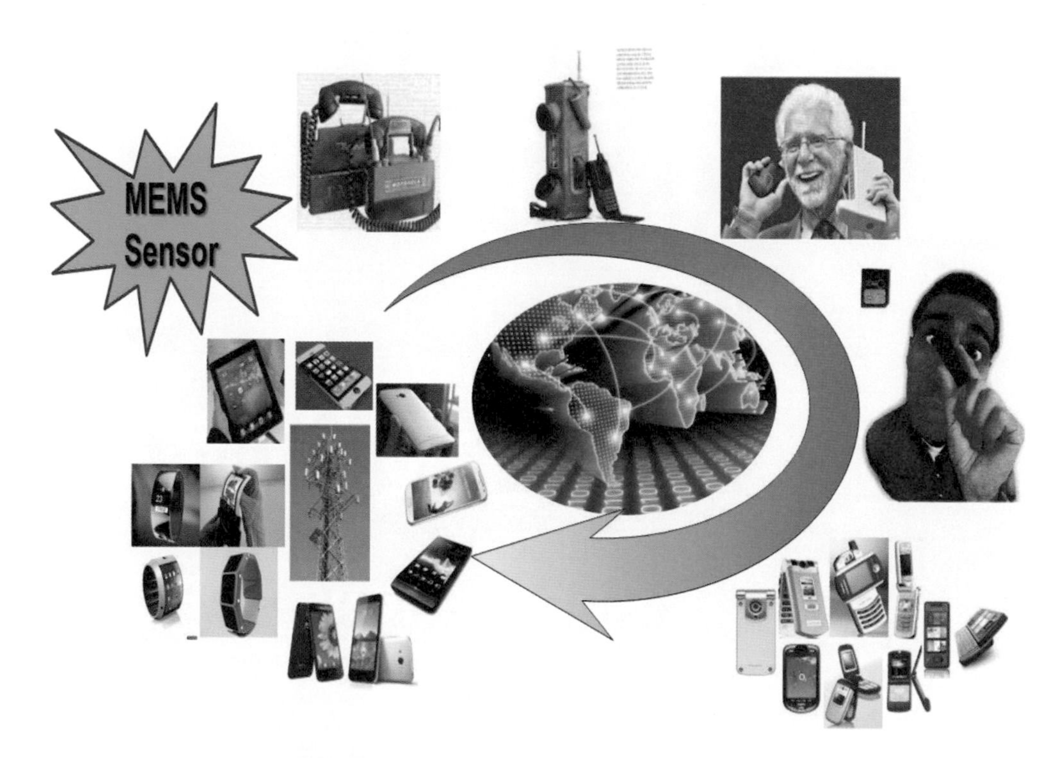

電子產品高度整合，3C 無法再區分應用了

As high integration, 3C is hard to identify applications anymore.

Application of PCBs

As industry transferred from analogous to digital, PCB manufacturing was stimulated to provide high volume production. Walk-man/ Calculator/ Computer/ Telecom. devices replaced by Radios/ TV/ Record players.

Before we used to define the PCB applications with 3C(Computer; Communication; Consumer) but now it′s hard to do the same way for too much hybrid products already.

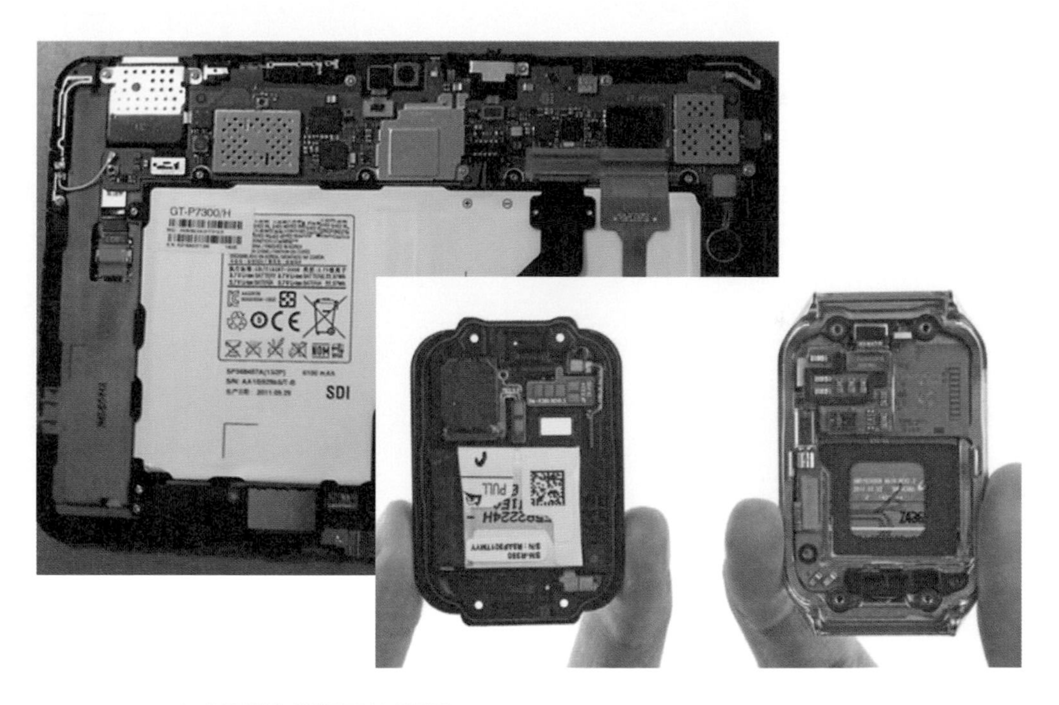

大小可攜式產品的應用 / Different size mobile applications.

4

電路板基礎製程簡介 — 硬板篇

Basic PCB Manufacturing Introduction (Rigid board section)

典型多層硬式電路板製程

一般電路板製程編排變化十分多樣，並不能以單一製程進行描述。但爲了便於作概括性陳述，僅以較典型的製程順序進行整理。

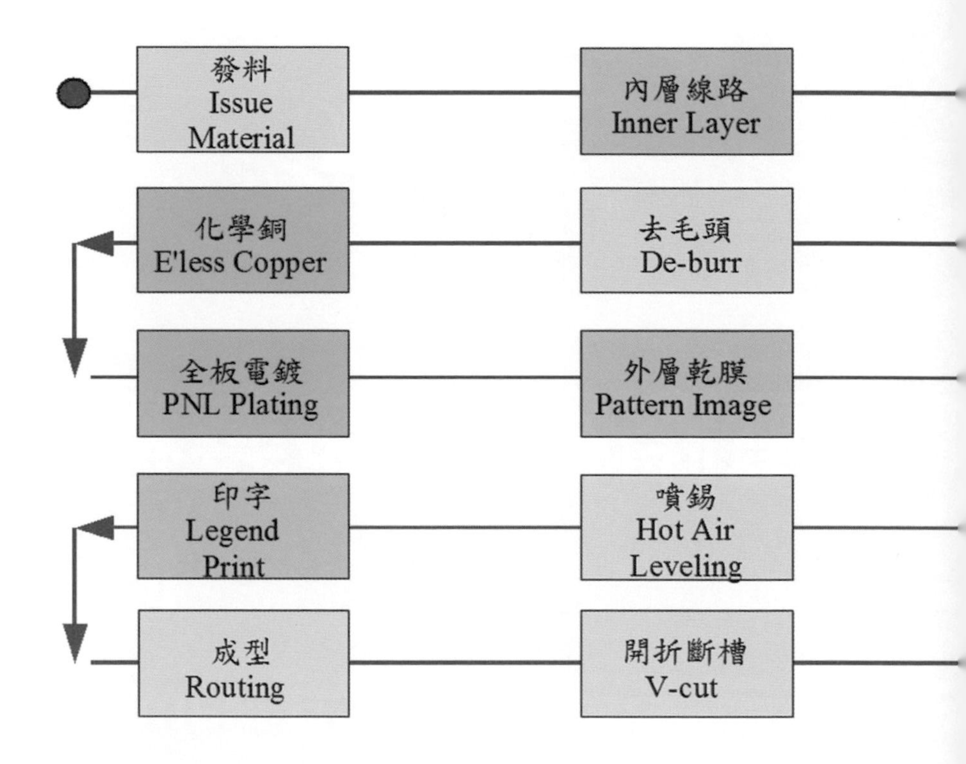

典型多層板製程 / Typical multilayer PCB processes.

Typical Rigid Multi-layer PCB Manufacturing Processes

Normal PCB manufacturing sequences are really flexible & can't applied with just one simple processes organization to cover all. But for easier describe this topic, the writer would like to apply a set of typical processes as the writing base.

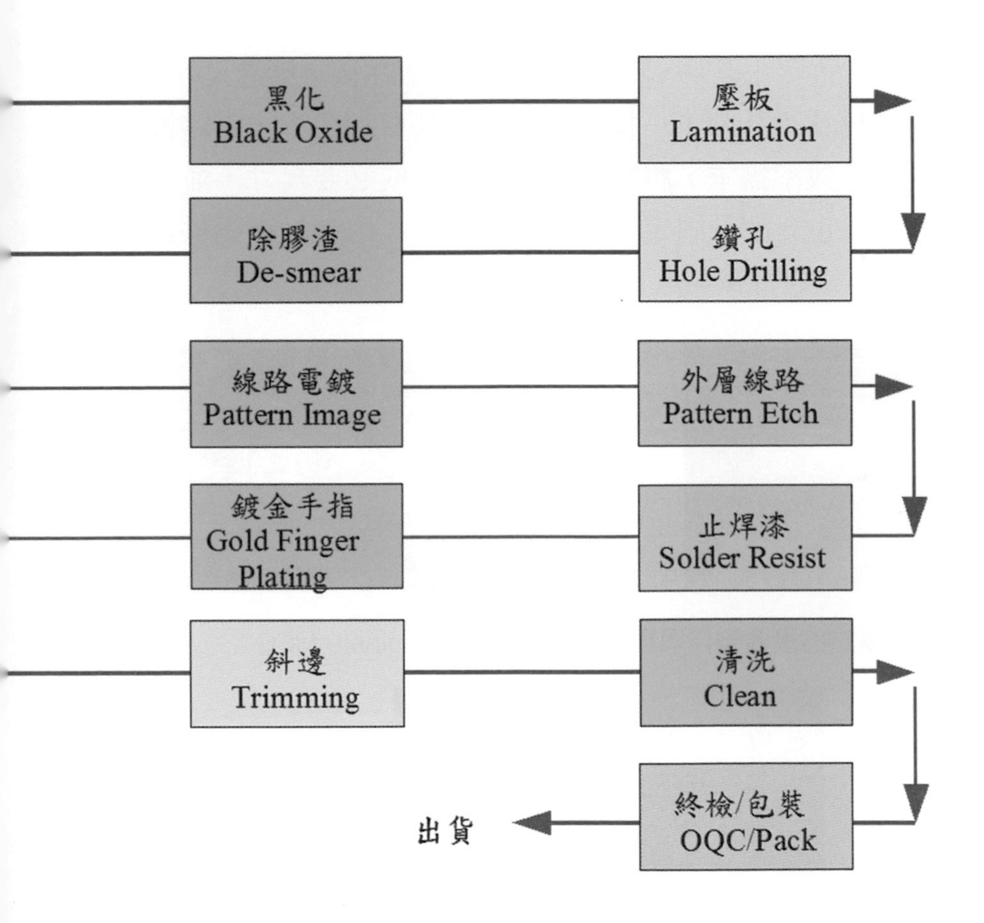

發料

製程說明

主要目的是爲了將大張電路板基材，裁切成一般電路板製造操作的尺寸。

工作內容

一般電路板基材，在基材製造廠商所製造的尺寸都非常大張。電路板業者會依據產品尺寸需要，將電路板設計配置在特定製造尺寸內。因爲各種電路板尺寸設計都不一樣，爲了提高基材利用率，電路板廠都會針對不同製造效率需求，將電路板基材切割成必要尺寸，並規劃出最佳操作效率。也有部分固定生產尺寸的板廠，由基材廠提供裁切完成的原料，而不是自行裁切。

電路板基材的壓合製作過程 / Copper Clad Laminate MFG.

電路板基材到廠尺寸 / Sheet Size for Cutting

Issue Material

Process description:

In process; operator cut sheets to working size.

General procedure:

Laminate were shipped in sheet sizes. Makers will divide them to fit working & operation, that will be based on design for increase material utilization. As different demands, PCB shops will do layout for maximize processing efficiency. Some shops are working with fixed size panels. They won′t do cutting themselves & ask vender doing those for them.

鋸床切割至生產尺寸 / Diamond Saw Cut to Working Size

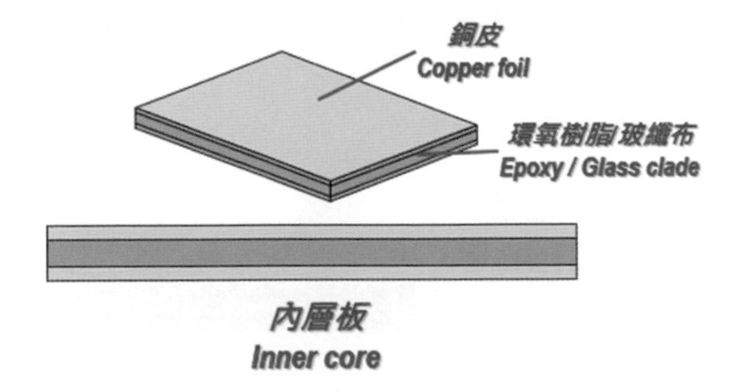

裁切尺寸符合生產需求 / Cutting size fit for manufacturing

光阻塗裝

製程說明

製作蝕刻阻抗層。

工作內容

電路板基材裁切後，必須針對不同設計，在基材上形成導電線路。對多層電路板而言，會有內部和外部線路之分；因此在製作內部線路時，會先將內部基材做適當前處理，作爲形成線路的”蝕刻阻抗層”塗裝表面。一般會先將基材表面脫脂、粗化和乾燥檢查，之後利用各種不同塗佈法，將蝕刻阻抗層塗佈在內層基板表面。

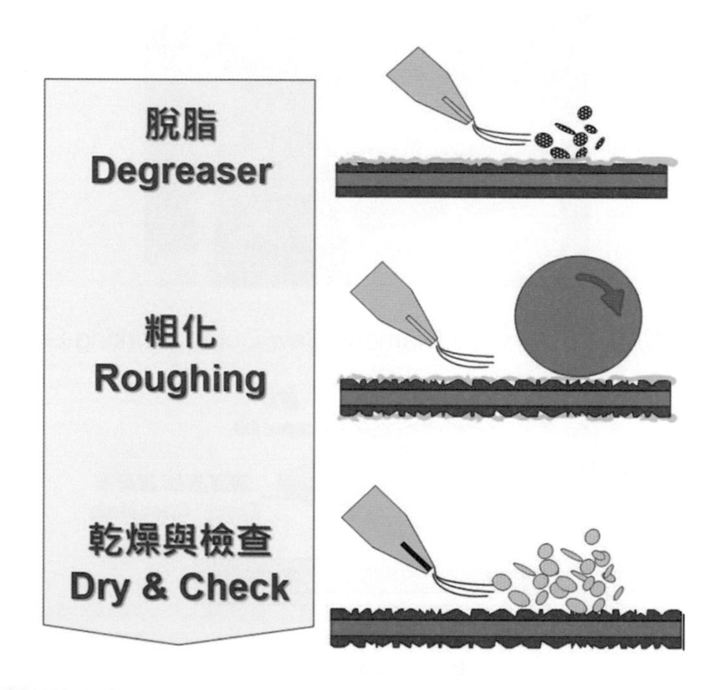

典型的影像轉移前處理 / Typical image transfer pretreatment

Etching Resist Coating

Process description:

To create a resist layers on inner-cores.

General procedure:

After sheets cutting, the working panels will create patterns required on laminates. For multi layer PCBs, the patterns will have both inner and outer layer. For inner layer patterns, makers will processing with some pretreatment, those including chemical & some mechanical steps. Then the laminates are ready for resist coating. The general procedures are degreasing first & then roughing surface with micro-etch; scrubbing or pumice & then drying the boards. Then apply resist with variety coating method; then the resist layer was there.

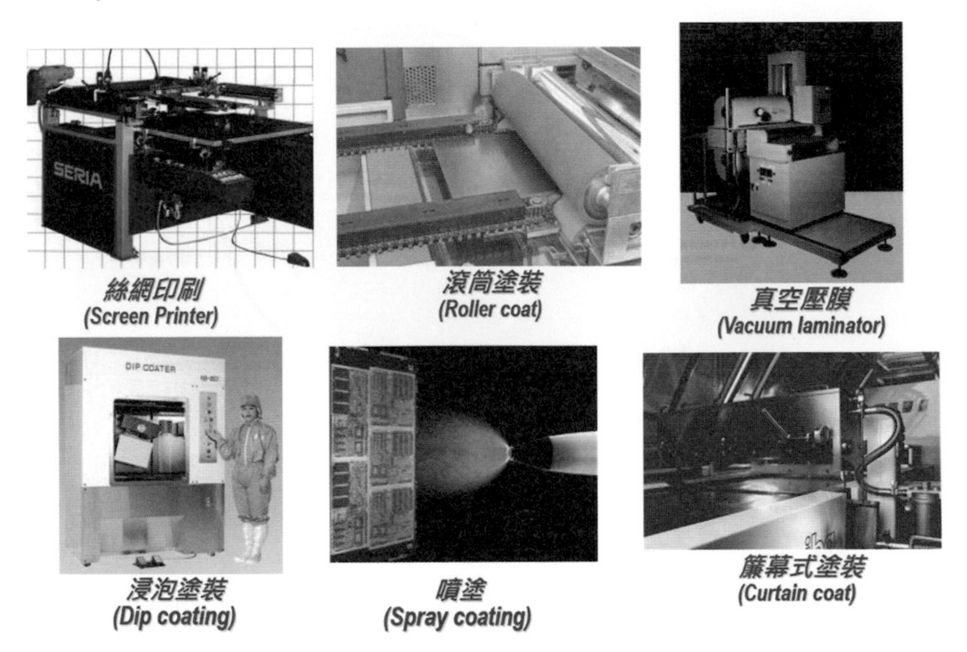

常見可用的電路板塗裝技術 / Typical PCB coating technology

乾膜壓合

製程說明

製作感光性蝕刻的阻抗層。

工作內容

目前業界普遍使用的製程物料，是所謂”乾膜光阻”，利用熱壓貼附在基材上。這只是塗裝技術的一種，因為製程簡單、操作方便、能自動化。對較高解析度電路板製作而言，環境潔淨程度十分重要，而這種乾膜貼附法，對環境敏感度及接受度都較佳，因此普遍用在電路板業界。當然，利用油墨塗裝、烘烤是較便宜的方法，但對於烘乾程序清潔度控制要求就需要小心，自動化是好解決方案。

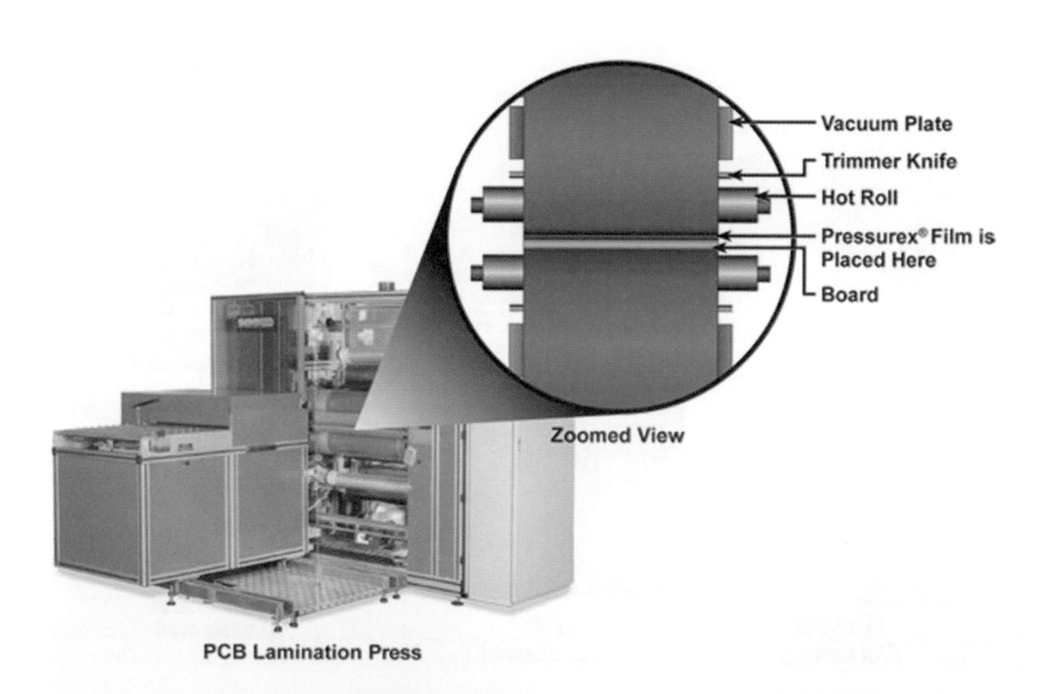

壓膜機 / Dry Film Laminator

Fry Film Lamination

Process description:

Create a photosensitive resist layer on boards.

General procedure:

The most popular photo resist used in PCB are so called "Dry Film". Applied with hot roll laminator, dry film can bond on boards as one of the coating methods. That is not so sensitive on cleanness & could be automated easily. It is now popular for fine line applications. Ink coating is cheaper for coating. But procedures are longer and sensitive on clean. Automation will be a good policy for that.

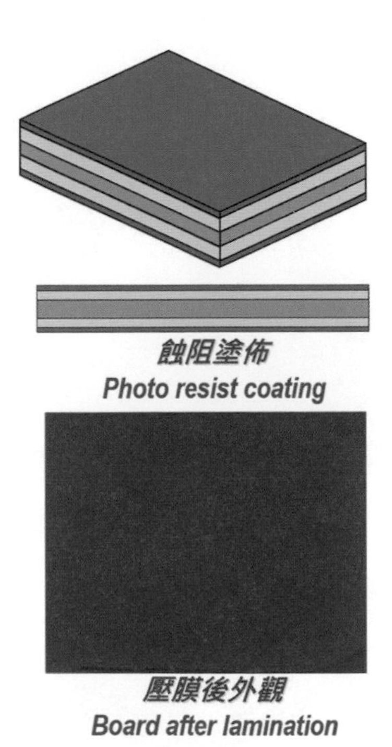

完成塗裝、停留後曝光 / Coating & stay some time then exposure

内層線路影像轉移

製程說明

利用曝光將底片影像轉移到板面。

工作內容

傳統電路板，尤指早期線路細緻度不高者，可用網版將線路直接印到板上。但由於線路逐年細緻化，傳統印刷已不能滿足電路細緻需求，而開始改用照相式“影像轉移技術”。所以在感光乾膜塗裝後，會用紫外曝光手段，將底片的線路影像轉到感光膜，再經顯影呈現出電路。近年因爲直接成像曝光設備與材料成熟普及，不少產品已轉用這類設備生產。

傳統與直接成像的曝光機 / Standard & DI exposure machine

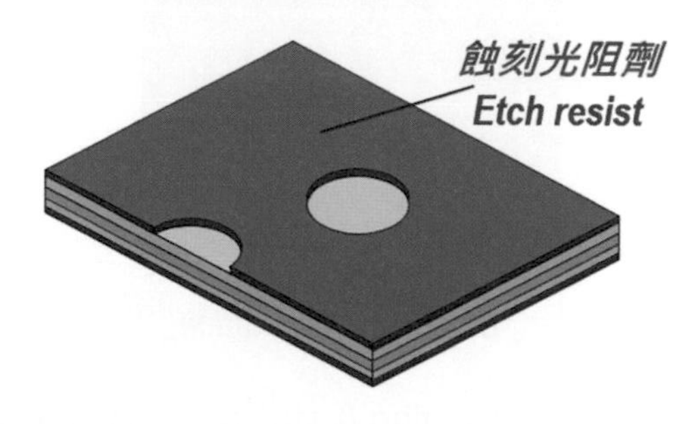

曝光顯影後的狀態 / After exposure & developing

Inner pattern Image Transfer

Process description:

Transfer artwork images from to boards.

General procedure:

Conventional PCB has loss & tolerate pattern. Screen-printing could be a suitable way for that. As pattern getting fine, screening won't work anymore. Makers applied image transfer as the solution. Using photo mask & exposure transfer images from artwork to boards. Then get in developing step, unexposed area will be removed from surface; then pattern come out. Recent years the DI systems & dedicate resist getting mature, some maker introduces those for mass production already.

電路板生產用的底片 / PCB use photo film.

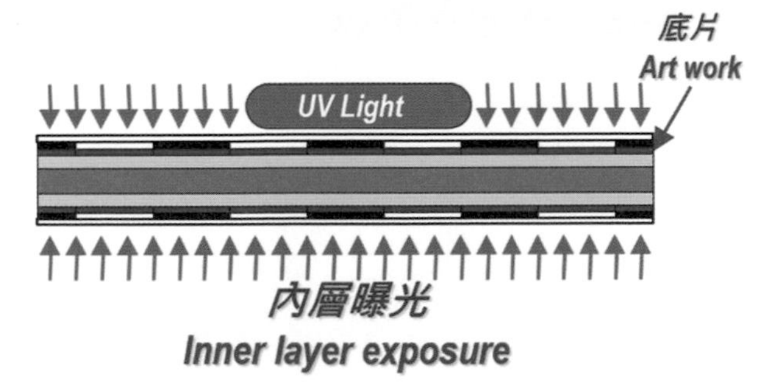

內層曝光 / Inner layer exposure.

機械力形成好的光阻劑側壁

製程說明

用藥液特性從事顯影、蝕刻、去膜工作。

工作內容

電路板對線路呈現，多使用濕式處理，利用藥水特性和機械噴流將線路顯影、蝕刻、乾膜剝除等完成。圖面呈現噴流機構噴出流體形式及影像膜反應現象。電路板線路處理，常使用水平設備，因此均勻度達成，必須要有機構設計才能改善。這類設備的噴流機構，必須考慮搖擺、噴灑角度、噴壓控制，來創造均勻反應環境，藉以達成全面反應一致性。

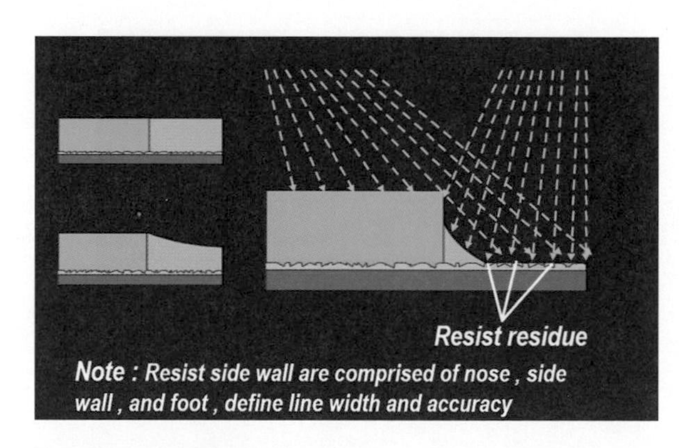

顯影過程光阻的清除狀況 / Photo resist under developing

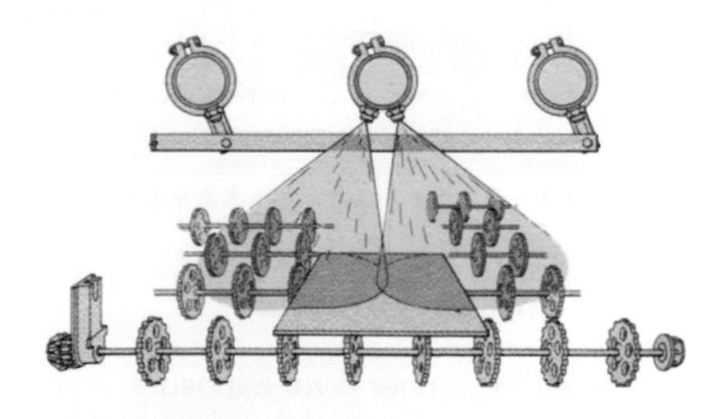

噴灑機構示意 / Spray pattern forming

Mechanical Force forming Fair Pattern Side-wall

Process description:

Do Developing; Etching & Stripping.

General procedure:

PCBs will pass through many wet processes for patterning. So chemical & fluid dynamic technology were applied for Developing; Etching & Stripping frequently. The diagram shows the spray pattern and resist interaction phenomenon. Generally, conveyor will be primary equipment, but reaction uniformity will be an issue. Makers have to aware of how to achieve a right spray pattern by nozzle oscillation /spray angle/pressure control. These efforts targeting on balancing reaction cross over boards.

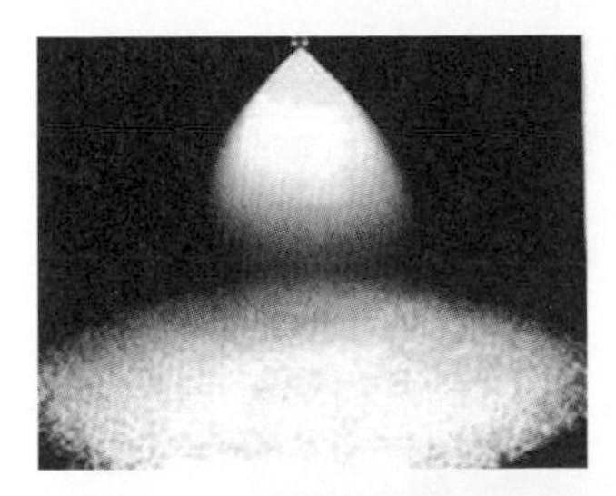

噴嘴噴灑的分佈 / Nozzle spray pattern

內層顯影
Developing

內層顯影 / Inner Layer Developing

顯影、蝕刻、去膜

製程說明

顯影、蝕刻、去膜的實際作業。

工作內容

一般多層電路板內層作業都採用顯影、蝕刻、去膜連線設備。因此處理設備的前段，會看到需要覆蓋的區域在顯影後呈現出來，而未覆蓋區域經過蝕刻段金屬會被蝕除。此時電路板會被送入去膜段，將殘餘感光膜完全去除，這是內層線路典型處理程序。圖面呈現的就是典型顯影、蝕刻、去膜線，一般業界統稱“DES Line”。

典型顯影、蝕刻、去膜線 / Typical DES Line

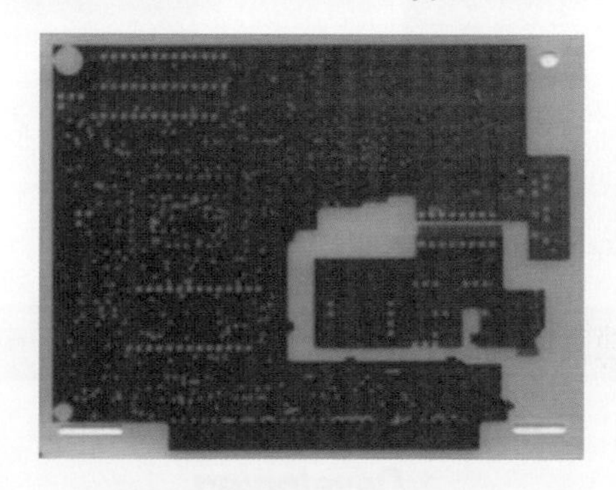

內層板經 DES 線後的狀態 / Core board after DES Line

Develop, Etch, Stripping Line

Process description:

The actual Developing/Etching/Stripping line′s operating procedures.

General procedure:

Inner layer wet processes were designed to Develop/Etch/Strip together subsequently. So that resist pattern will come out first after developing, non-exposure areas will be removed & copper exposed. Exposed copper will then be etched off through the etching chamber & get into the stripping chamber. Stripping chamber will remove all the etching resist remaining and leave the copper pattern only. This is the typical inner layer wet process line. The picture below is a typical so called “DES Line”.

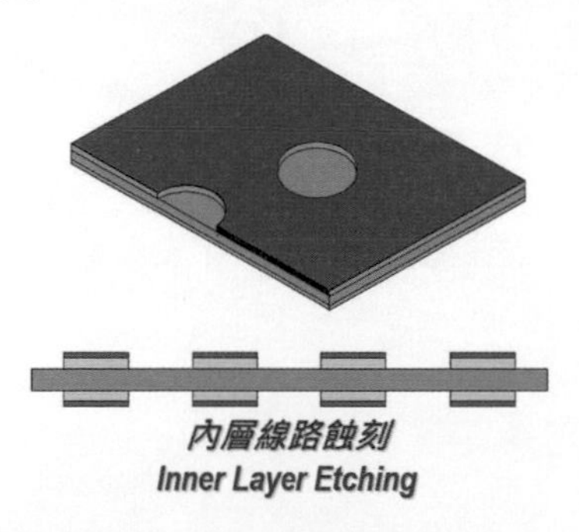

內層蝕刻完畢剝膜前 / After etching & before stripping

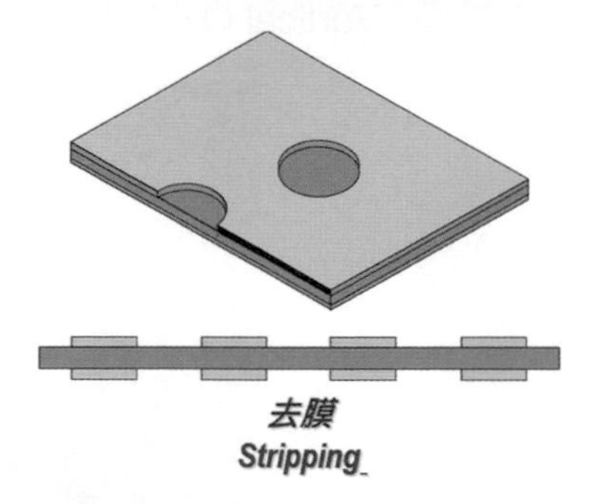

剝膜完成內層線路呈現 / Resist stripped & pattern made.

表面粗化(黑化)

製程說明

粗化線路表面，加強金屬與樹脂的結合力。

工作內容

內層線路呈現後，銅面仍相對平整。金屬與非金屬結合力，若僅靠樹脂與平坦金屬鍵結，很難有良好結合力。因此電路板製作，會將銅面做適當粗度，強化樹脂與金屬結合力。常見方法有所謂“棕化”與“黑化”，作業是利用化學品與金屬面作用，以成長或蝕刻，將銅面打粗或做成毛面，藉以完成粗化。圖面呈現的是黑化生產線及黑化後表面狀況。

垂直黑化設備 / Vertical Oxide Equipment

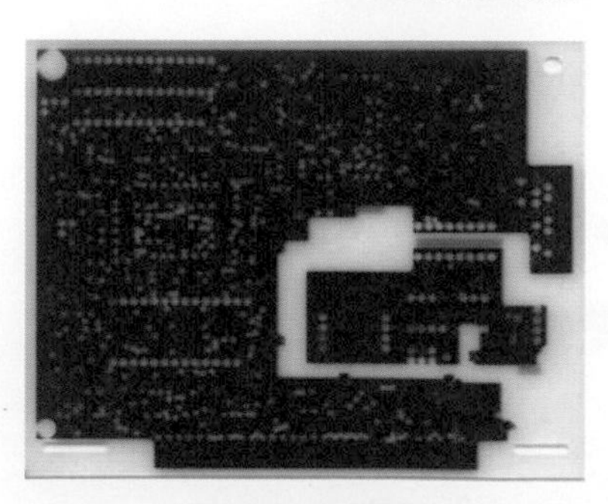

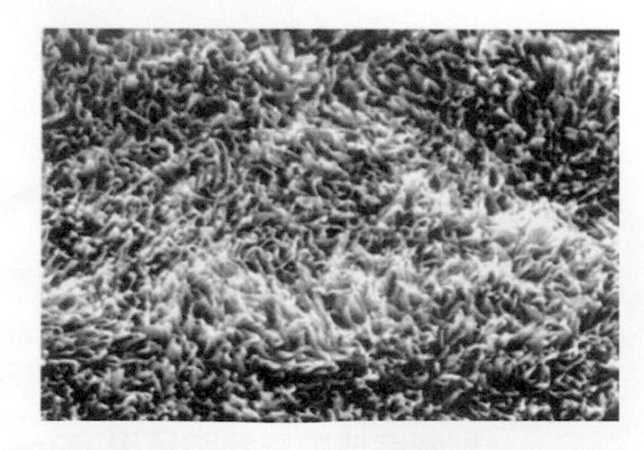

黑化後電路板外觀 / Boards after Oxide Treatment

Surface roughing (Black Oxide)

Process description:

Create roughness on copper for better bonding.

General procedure:

After pattern etching, the copper surface was still smooth and not fit for resin bonding. Bonding between metal and non-metal surface depend on mechanical force; then creating copper roughness is necessary. Most popular roughening processes used in the PCB industry are “Black” & “Brown” oxide. The process uses chemical solution for anchoring. Base on that topography, the bonding strength can be improved. Below pictures typical oxide lines and the board′s surface after oxide.

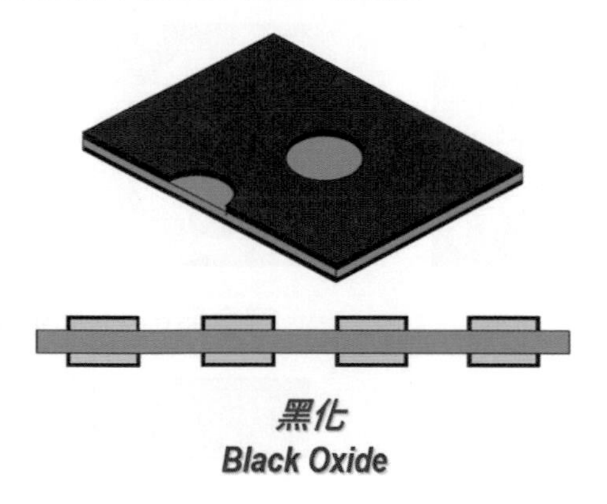

黑氧化與棕化處理 / Black oxide and brown oxide.

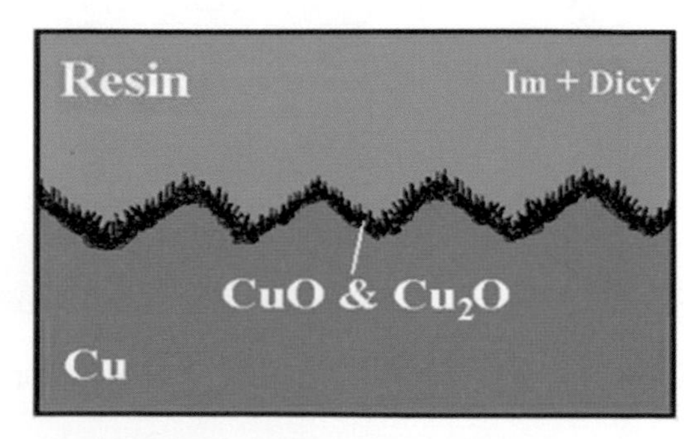

鉚合與固定程序

製程說明

鉚合電路板，固定板間相對位置。

工作內容

電路板設計厚度不一，因此鉚合不見得適合所有需求。壓合過程因爲樹脂流動，基材會有一定壓縮量。如果壓縮量超過鉚釘長度界線，鉚釘也會因壓縮變形，這就失去固定功能，反而因鉚釘變形造成錯位。因此對較高層次基板，製造商常採用插梢式壓合機構，圖面呈現的是鉚釘外型及鉚合結構，另一圖則呈現插梢式壓合結構 (Pin Lamination)。

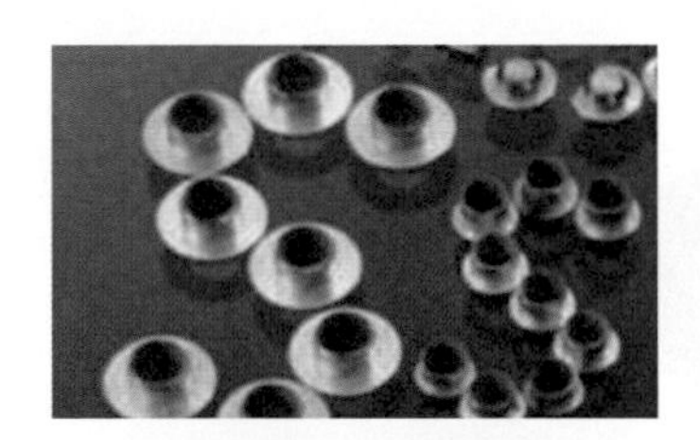

鉚釘 / Rivets

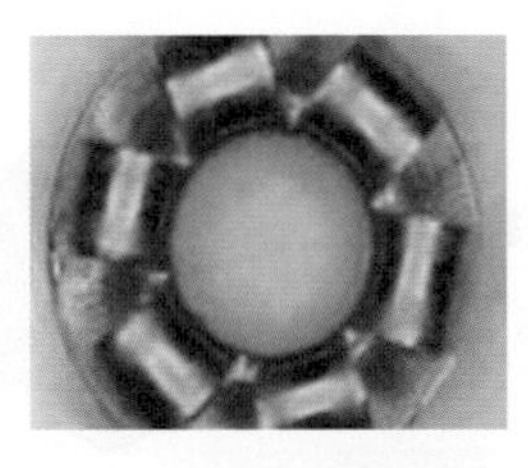

鉚合品質 / Riveting quality

鉚釘機 / Revit machine

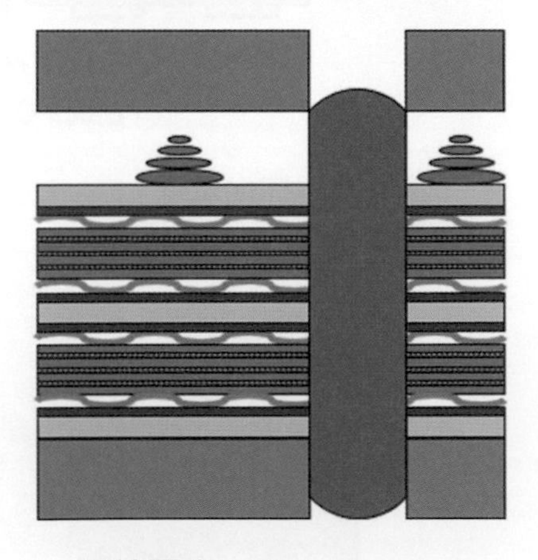

插梢壓合 / Pin lamination

Riveting and Booking Procedure

Process description:

Riveting PCBs to keep registration.

General procedure:

PCB thickness are flexible and riveting may not fit all design. When do the hot press, resin will flow to fill spaces. Then total thickness will be reduced under resin flow and pressed. While thickness reduced to less than rivets height, then will be bended that might initiate shift. For high layer count boards, it's hard to use rivets for thickness decreased too much. Then makers try to use "Pin Lamination" . The diagram shows the riveting and examples of good & bad riveting. The other diagram shows the pin lamination structure.

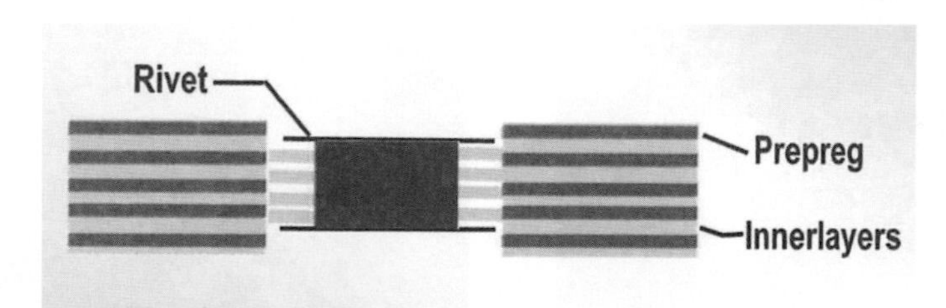

良好的鉚合 / Good Riveting

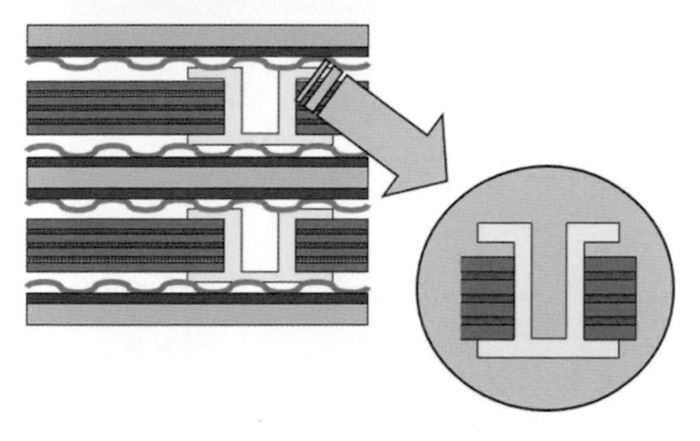

不良的鉚合 / Bad Riveting

壓合堆疊

製程說明

組合多層線路進行結構堆疊。

工作內容

內層板經粗化，必須透過疊合加膠片程序才能將電路板完全組合。一般組合會先將內層板依排序，利用對位機制，如：對位孔等，將電路板相對位置固定。常用的固定模式，是金屬釘鉚合，也有部分會採用預貼合模式作業。組合後的中間品稱爲“Book”，將這些“Book”再疊入樹脂片及上下銅皮，就可以進入壓合，這個疊合稱爲“Lay-Up”。

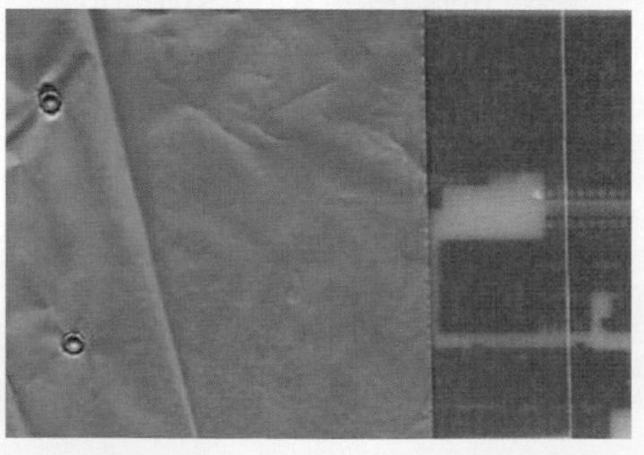

預貼合與疊合後外觀 / Pre Lam & baords after Lay Up

Lamination Lay-Up

Process description:

Organize the board sequence & stacking.

General procedure:

After oxide treatment, inner layer will be stacked as design. While stacking, operator will slide in bond sheets. The sheets so called "Prepreg" are kinds of semi cured glass clad carry sheet. The alignment will depend on tooling holes or other targets. Makers will rely on these to keep registration between layers. The production procedure was called "Mass Lam". Riveted inner layers were called "Booking" also. After booking, operator will lay copper foil/prepreg and add separation sheet on a carrier. The whole procedure was so called "Lay Up".

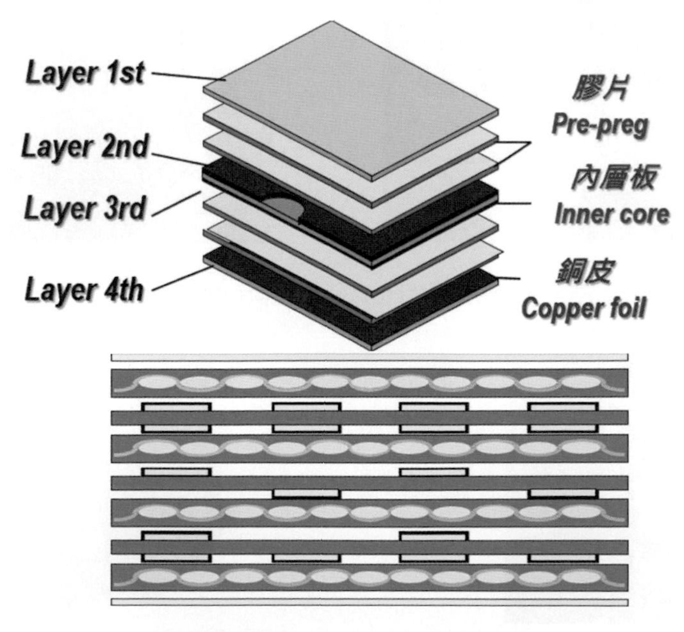

鉚合與疊合 /Booking & Lay-Up

熱壓合

製程說明

熱熔樹脂片，以眞空、膠流動將內層板填充結合並固化。

工作內容

電路板疊合後，可將疊合板送入眞空熱壓機，利用機械提供的熱能，將樹脂片融熔，藉以黏合基板並填充空隙。空隙填充後，繼續利用高熱使樹脂完成聚合。爲了使熱壓合傳熱程序均勻，目前壓合機設計多使用熱油加熱循環系統，如圖所示。在眞空壓合機旁會設置冷壓機，也會有傳送台車的配置，目的就是爲了方便移動基板，並在完成熱壓後利用冷壓冷卻基板，以便下料及下一製程後續處理。

冷熱壓機系統 / Hot & cold press system

熱油爐 / Hot Oil Burner

Hot Press-Lamination

Process description:

Use hot oil heating system to melt the resin then apply vacuum chamber and pressure to fill up empty spaces and then cure the resin in the same time.

General procedure:

After lay-up, material will be sent into a hot press vacuum chamber. After heating up, resin's viscosity will reduce and fill up spaces between layers. After filling up all the spaces, the press will keep hot. Most of the hot presses are designed with hot oil heating systems with circulation to heat up the press. As the picture shows, a cold press & transformation platform followed. The functions of these are moving boards between devices and cooling down PCBs for next stage operation.

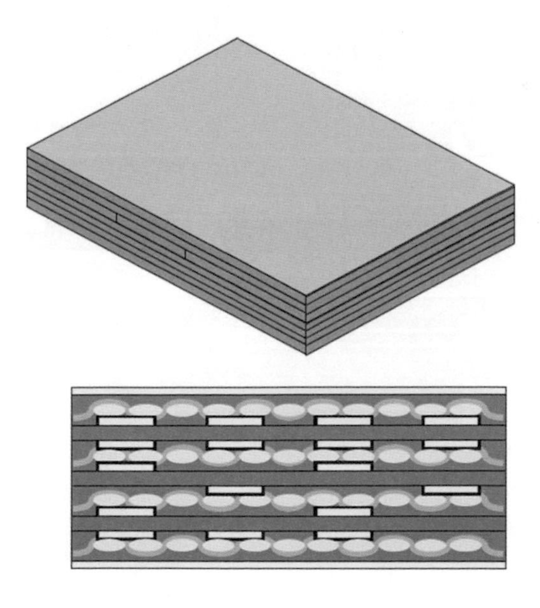

熱壓合後 / After hot press (Lamination)

油壓式熱壓機構

製程說明

大量生產電路板最常用的疊合模式及壓合設備模式。

工作內容

電路板疊合作業，是多片堆疊。電路板間爲了平整性，都會疊入鏡面鋼板，在堆疊頂與底部，會加上緩衝墊或牛皮紙，藉以均勻壓力並減緩傳熱，促進加熱均勻度。壓合過程，最常使用的機械機構是“油壓式”熱壓機。樹脂受壓受熱，會因爲受壓與熱熔而向四邊推擠，經此過程完成層間黏合和空隙塡充。因此一般疊合上下面銅皮尺寸，都會大於電路板，以防止樹脂外流造成鋼板及機械污染。同時樹脂設計和電路板線型設計，也都會將樹脂流動模態作適當設計和考慮。

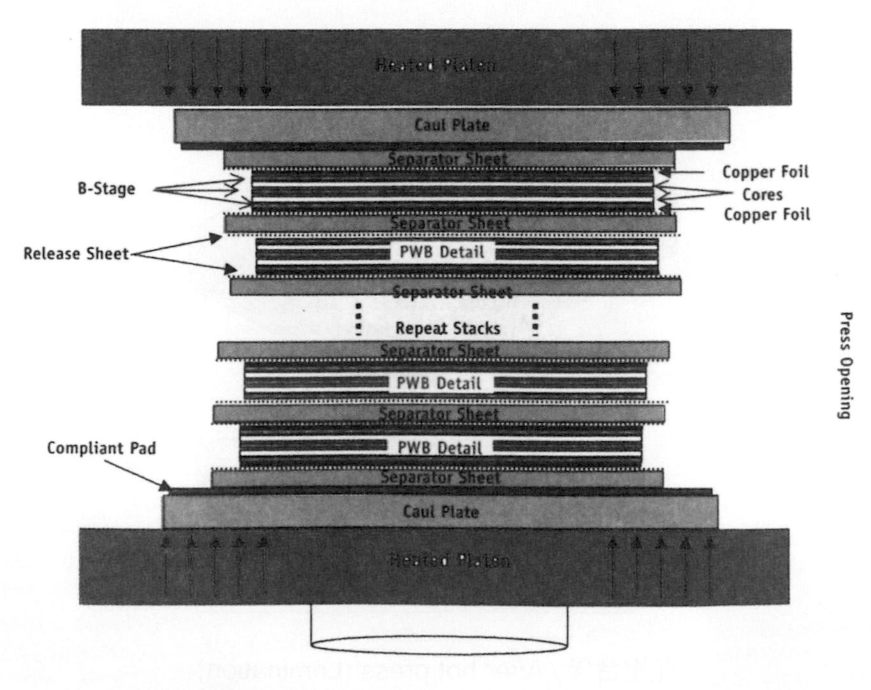

整體熱壓疊合機構 / Lay Up Structure

Hydraulic Hot Press Structure

Process description:

Architecture lay up & hot press.

General procedure:

Mass lamination isn't single board operation but rather multi panels in real practice. For guarantee flatness of the boards, stainless separating sheets are used as the media. Craft paper was stacked on top and bottom of boards as cushion and thermal buffer, to unify heat transfer and pressure. Hydraulic type was the most popular hot press used for lamination in the world. Because resin will flow and fill up space during heating and pressing, for prevent separating sheets been contaminated, then copper foil size will usually be larger than board size. Resin flow can also be controlled by resin design & pattern dam design.

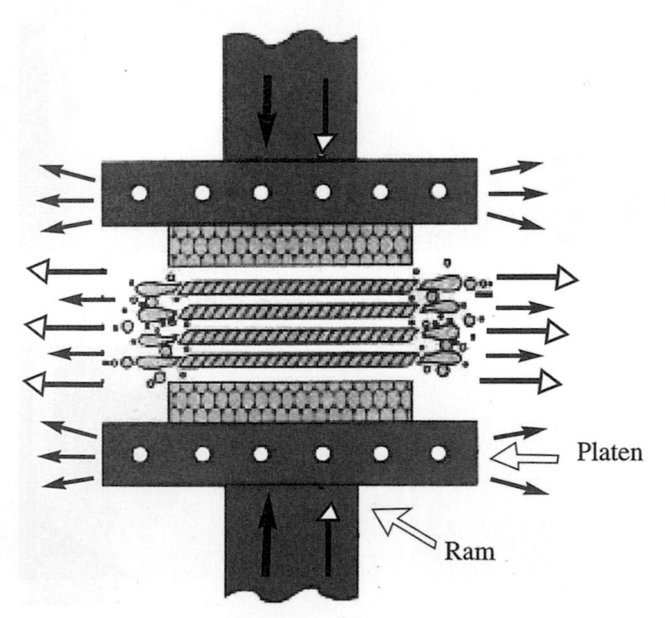

樹脂側向擠出機構 / Resin Squeeze Out

壓板後處理

製程說明

移轉座標，將座標系統轉移到電路板表面。

工作內容

電路板壓合後，所有內層線路已被銅皮覆蓋，因此所有座標位置無法在表面看到。如此後製程沒有基準座標依循，必須有座標移轉機制轉換，業者採用的移轉作法為“鑽靶”。內層板在製作內層線路時，同時製作座標參考區。經過不同讀靶程序，將靶區利用鑽孔打出，給後續製程用。此時板面一片平坦，沒有任何記號可辨認，為了後續管理追蹤及製作便利，必須在板面打辨識碼藉以辨識。圖面所示為打號碼機的號碼頭，有部分產品已經採用雷射打標機取代。

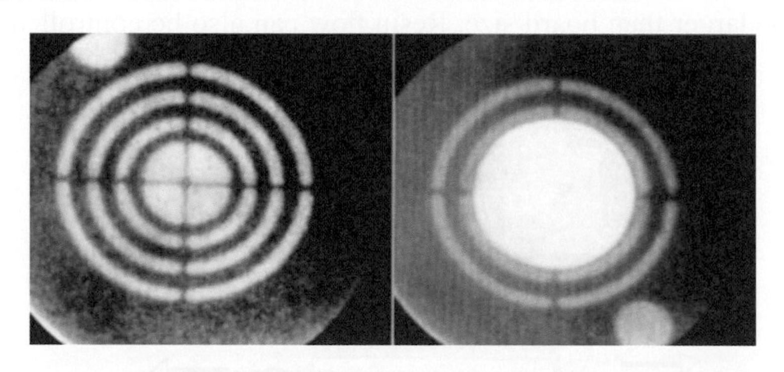

同心圓與靶標孔 / Co-circle & Target Hole

板面打標 / Marking

Lamination Post Treatment

Process description:

Transfer inner pattern targets to outer layer.

General procedure:

After hot press, all the patterns are masked by the copper surface. Then there are no reference targets on the surface. Makers will use so called "Target Drill" to transfer inner targets to outer layer.

The pictures shown below are target drill machine. After target searching and drilling, inner layer targets will transfer to surface. Because it is hard to distinguish boards apart at this point, for management concerns we stamp numbers on the boards to tell. The picture shown here is a marking head for number marking and some maker now use laser mark machine as replacement.

X 光自動鑽靶機 / X-Ray Automatic Target Drill Machine

機械鑽孔

製程說明

利用機械鑽孔產生金屬層間連通管道。

工作內容

完成壓板後，內層線路層間沒有連通，無法達成層間連通，需要以鑽孔方式將層間打通，目前普遍使用的鑽孔機是數位控制型多軸鑽孔機。這種鑽孔機以數位驅動、螺桿滑軌定位，運動路徑及停止位置，利用內藏光學尺達到高精準度。一般鑽孔都採多片作業，爲了完成高精準度鑽孔，多片疊合會在上方覆蓋鋁板，藉以加強散熱及保護電路板，底部則用犧牲墊板，保證鑽孔穿透性及避免傷害鑽孔機檯面。

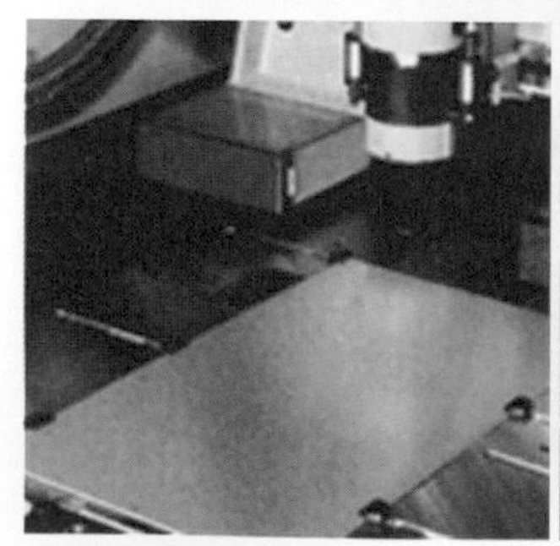

板挾持機構
Board Clamping

趨動馬達
X-Axis Motor

光學尺
Optical Meter

Mechanical Drilling

Process description:

Create connection between metal layers.

General procedure:

After hot press, non interconnection between layers, we have to create through holes to achieve connection. Makers use CNC type mechine to the drilling as shown in the picture.

The drill table was driven by ball screw and linear guidway control the drill location. The drill registration was base on NC program and built-in optical meters guiding. General drilling operations will be stacking panel mode. To get high accuracy, top of stack will cover with an Al sheet for protection & improve heat dissipation. A back up panel will be used at the bottom of the stack as a sacrificed board. That can guarantee the holes were drilled through completely, and no table damaging/drill bits breaking/ position shift issues occur.

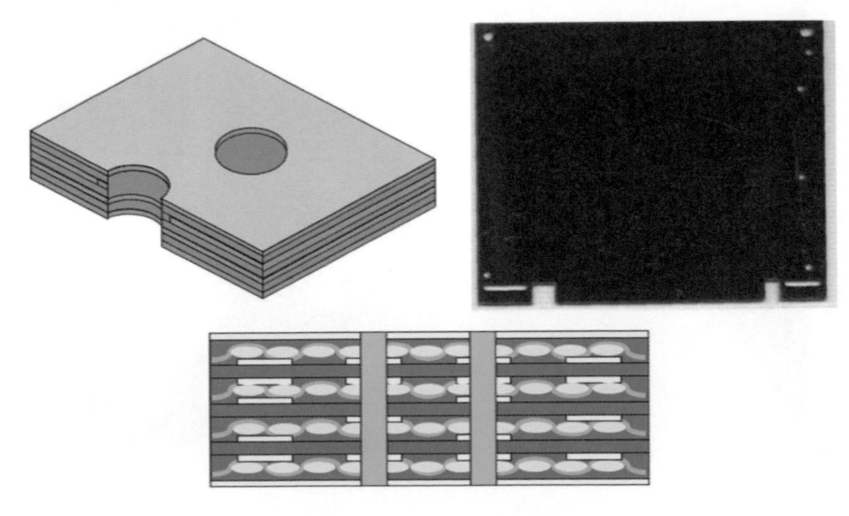

機械鑽孔 / Mechanical Drill

氣墊式鑽軸

製程說明

鑽孔機挾持及鑽軸機構。

工作內容

電路板鑽孔會有不同尺寸需求，且電路板是複合材料，採用的鑽孔操作也與一般金屬加工不同。爲了切割樹脂、銅皮、玻纖三者，電路板鑽孔加工採用高速旋轉加工法，目前常見加工轉速爲 16、20 萬 RPM。高速旋轉使用的軸承，無法採用傳統滾珠軸承，必須採用特殊氣式軸承，因此機台成本與所用軸承、轉速有相當關聯。軸承前端會加上鑽針挾持夾頭，有雙瓣、三瓣、多瓣等不同設計。作業利用伸縮及夾頭斜度控制鑽針抓取，因此夾頭眞圓度十分重，操作保養清潔，也會影響位置精度及孔壁品質。

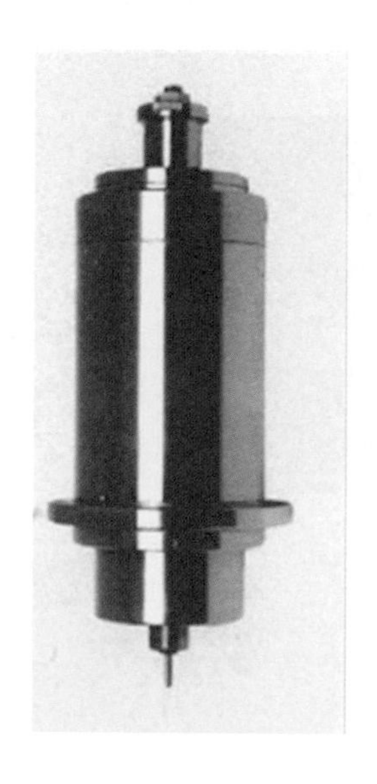

鑽軸組 / Spindle Sets

Spindle for Drilling (with Air Bearing)

Process description:

Describe spindles and drill bit clamp structure.

General procedure:

One board with different holes sizes were normal & PCB made by composite materials. The drilling will be different from metal base material.

For cutting resin; copper; glass fiber all together, drilling need high speed operating. 80,000~300,000 RPM equipment are available on market. Air bearing was applied on spindle but not ball bearing. Machine cost will be highly related to the bearing applied.

Handling drill bits and replacement will depends on clamps. There are several types of clamps used on machine. Under running, clamps will move in and out to replace the bits. Clamps orifice roundness & clean will be critical for drilling quality. Maintain are very critical to achieve good drilling quality.

鑽針夾頭 / Drill Bits & Clamps

鑽孔鑽針的排屑

製程說明

鑽針應用及良好鑽孔排屑的必要性。

工作內容

鑽孔過程中會產生高熱，但電路板加工多數採用乾式鑽孔，不會加任何冷卻油或水，因此排熱、排屑就是鑽孔品質重要因子。圖面所示排屑現象，是鑽孔排屑動態畫面，一般鑽孔機爲了達成位置精準度及良好孔壁品質，都設計了不同的所謂“壓力腳”。壓力腳功能主要是固定電路板，同時提供高速氣流的機構，藉以保證鑽孔排屑良好，同時強化冷卻能力來改善鑽孔品質。

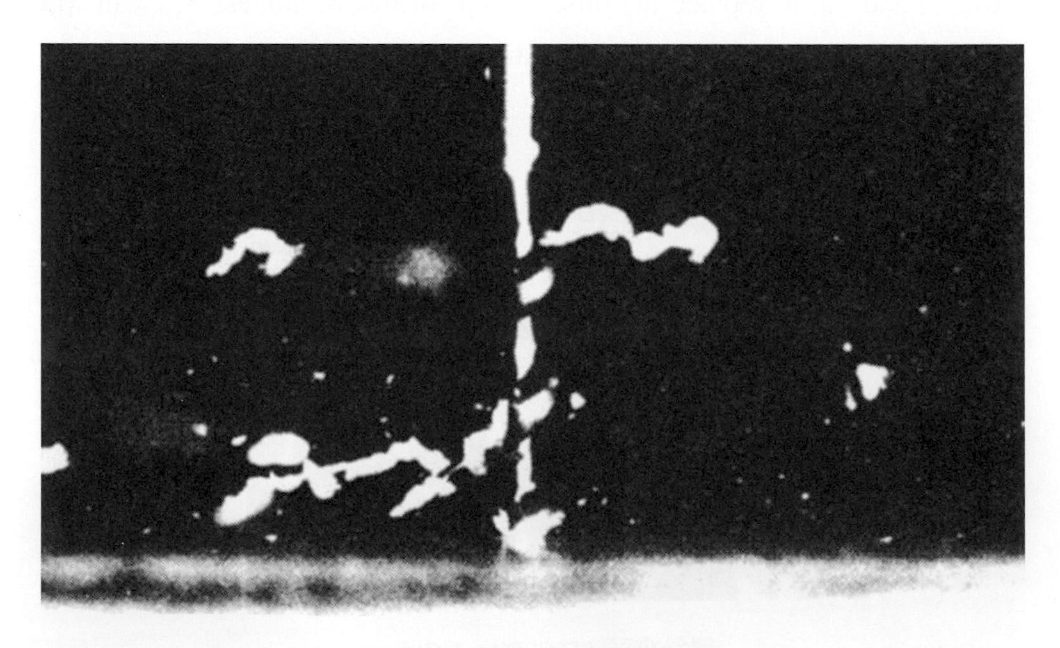

鑽孔排屑量 / Chip Load

Drilling Bits & Chip-loading

Process description:

Chip load on working drill bits will be important.

General procedure:

PCB drilling will generate enormous amount of heat, as it was operated withoout water or oil cooling. How to improve the heat dissipation and chip disposal is very important. The picture showed the dynamics of chip disposal.

PCB drill spindles have so called "Pressure Foot" mechanism as shown on the diagram. The main function of that was provide sufficient airflow to cool down the drill bit and dragged out chips by the air. In the same time air will spread heat through aluminum sheet also. With these features, drilling quality can be improved significantly.

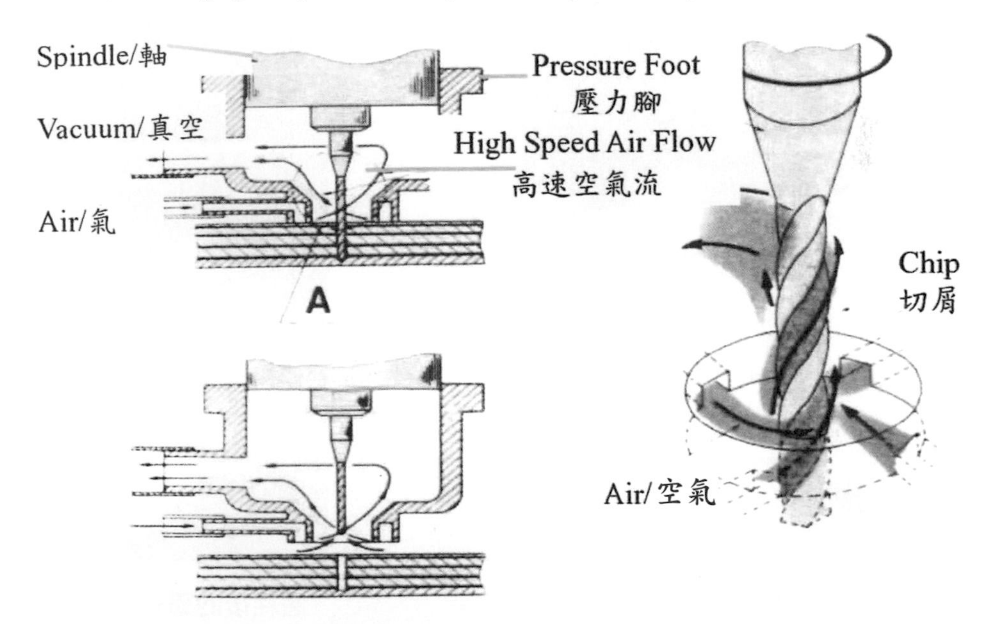

鑽軸頭的作業結構狀態 / Drill Head Operating Structure

鑽針研磨及排放

製程說明

鑽針的運用。

工作內容

鑽針是電路板鑽孔主要耗材，極度影響電路板鑽孔成本，因此多數大孔徑鑽針都會再研磨，下圖所示，是鑽針磨耗後圖像及研磨機機構。鑽孔程序會使用鑽針套環，主要目的是爲了能控制鑽針刀面韌長，促使鑽孔機抓取時一直保持一定突出長度，這才能保證鑽孔作業中不鑽不透或過深、斷針問題。因此鑽針處理還包含鑽針套環使用及萬用鑽盤設置，許多大廠將此作業分開獨立運作，可促使鑽孔程序順暢。

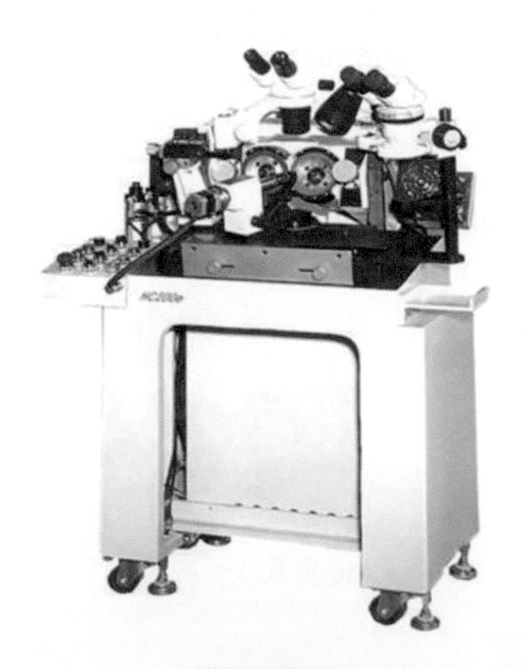

鑽針研磨機

Drill Bit Re-sharping Machine

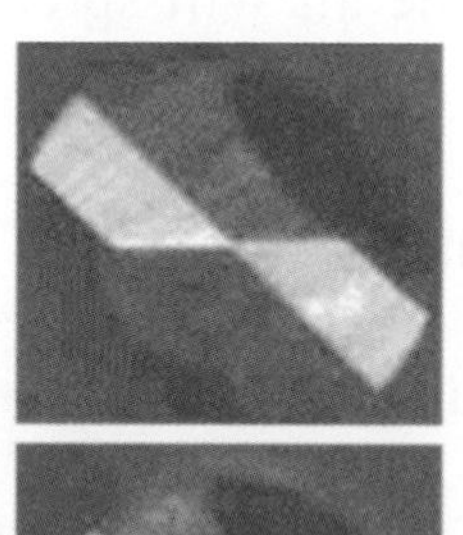

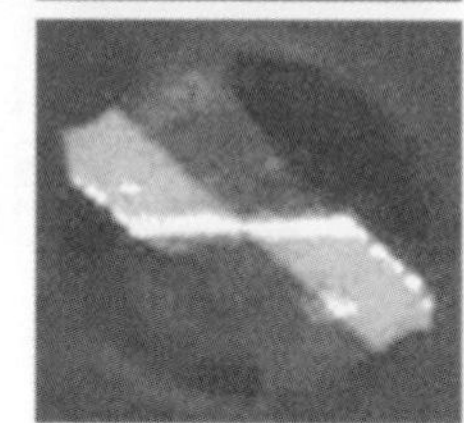

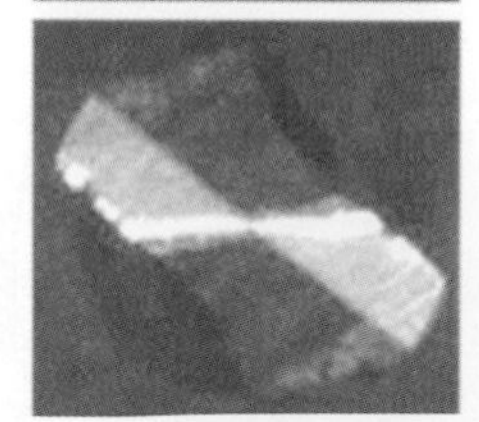

磨耗後的鑽針

Wear Out Drill Bits

Drilling Bits Re-sharp & Setting

Process description:

Drill bits application.

General procedure:

Drill bits are main cost of drilling, reuse & re-sharp were main issue for operation. Big size drill bits re-sharp is routine for MFG. The typical re-sharp machine show as below.

New drill bits or re-sharp ones will be loaded with depth control rings then set on cartridge. Rings loading function are depth-control, for guarantee drill bits pass through boards but won't damage the table. The operation will include rings loading / set into universal cartridge & load on machine. Some high volume production will have dedicated group focus on that. Then can make operation smoothly.

鑽針套環 / Ring Setting

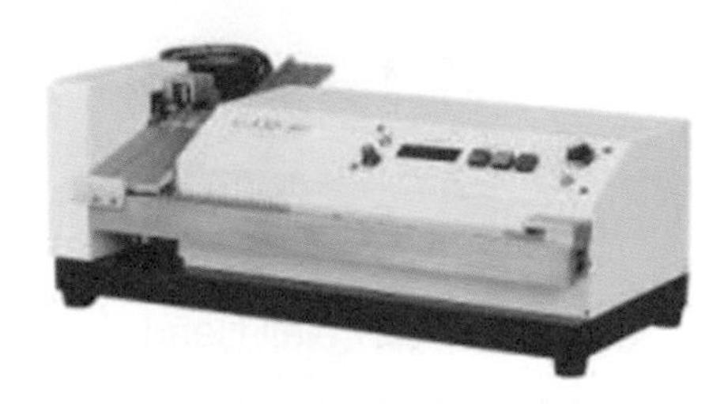

鑽針套環機 / Ring Setting Machine

萬用鑽針鑽盤 / Drilling Bits Cartridge

釘頭缺點及對位偏差 (鑽孔及電鍍後)

製程說明

鑽孔品質確認。

工作內容

鑽孔後，必須確認位置恰當、無漏鑽、鑽偏、鑽破或內部撕裂缺點。業界常用檢查設備，為 X 光透視檢驗機，對電路板鑽孔位置作抽樣檢查，藉以確認鑽孔區域沒有超越銅墊限制範圍。孔內品質方面，則須確認無過度撕裂現象。如果有銅層與樹脂撕裂現象，則會在電鍍後呈現所謂“釘頭型缺點”，過大釘頭會形成未來品質信賴度問題，因此多數產品都對此種缺點作出規格定義。

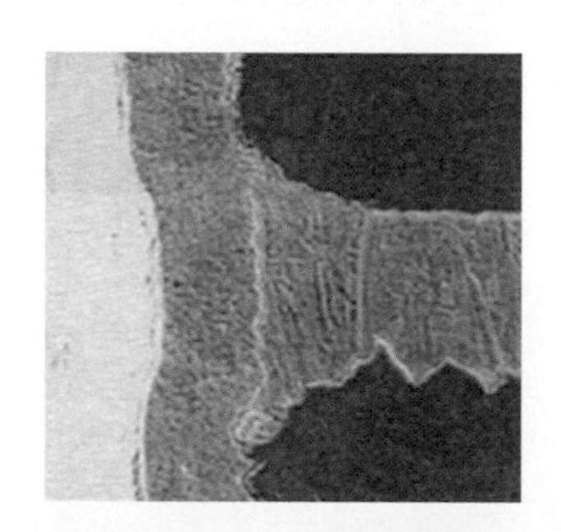

釘頭型缺點 / Nail Head

X 光檢查機 / X-Ray Inspector

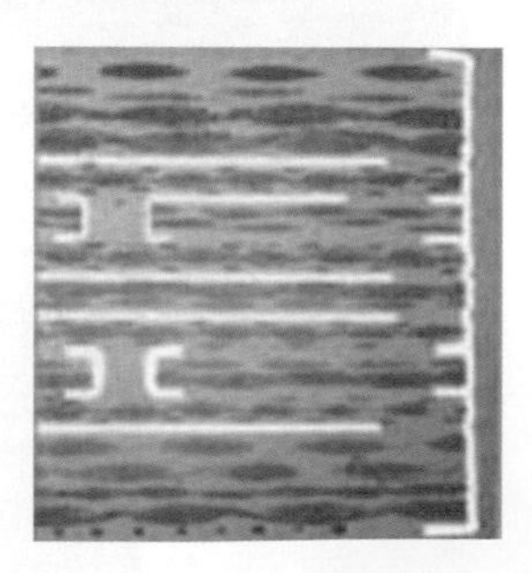

品質良好的孔 / Quality Passed Hole

Nail Head & Miss Alignment(After Drilling and Plating)

Process description:

Identify drilling defects.

General procedure:

Drilling quality mainly focuses on registration / miss drill / hole's wall tear and so on. X-ray inspection is popular for in process quality control.

X-ray used in alignment check & hole to copper pad miss-registration. Hole's wall check will focus on inner-layer tear issues. Resin & copper interface was torn off severely, then "Nail Head" will appear after plating. Most products will define the acceptance Spec. for this issue.

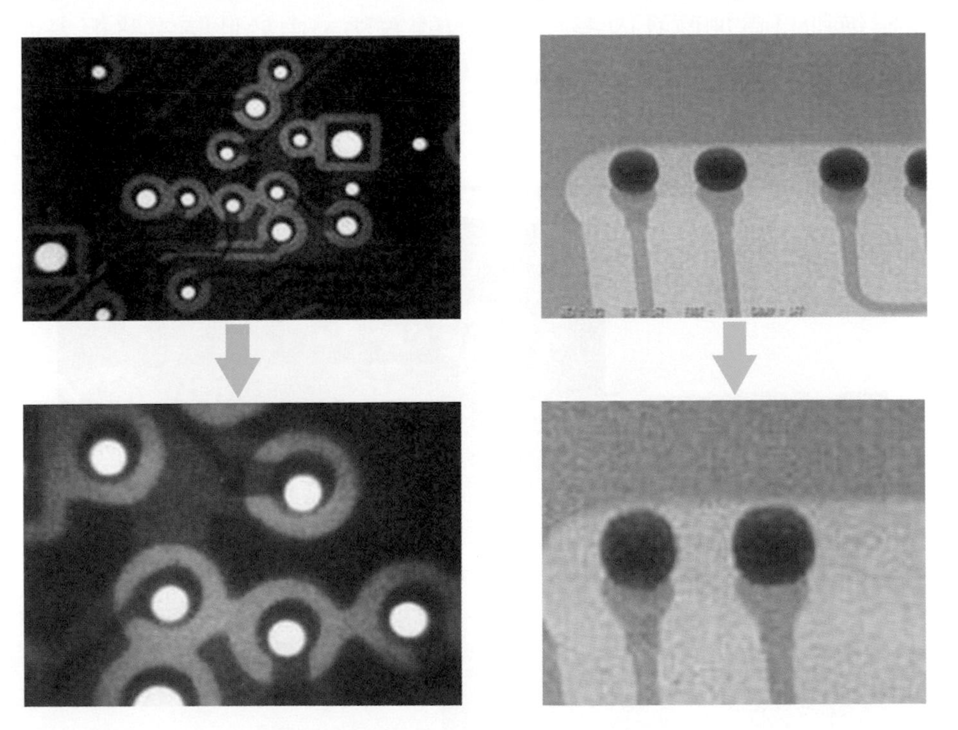

鑽孔位偏缺點 / Rigistration Defects

鍍通孔製程

製程說明

連接層間的銅線路。

工作內容

鑽孔完成後，層間電路並未導通，必須在孔壁形成導通層以連通線路，業界稱爲“PTH 製程”，主要工序包括：除膠渣、化學銅和電鍍銅三個程序。除膠渣，主要目的是清除孔中產生的殘渣，可能對內部線路導通造成困擾。化學銅主要目的，是清潔孔壁形成導通銅薄膜，促使電路板後續電鍍有良好導電力，近來也有廠商使用其它導體，如：石墨或導電高分子作爲介質再電鍍。這類製程被稱爲“直接電鍍”。全板電鍍則是爲加厚孔內銅厚度採取的方法，由於此時電路板表面線路尚未形成，因此這種程序被稱爲“全板電鍍”。

化學銅與電鍍線設備 / Electroless Copper & Plating Equipment

Plating-Through Hole Process

Process description:

Creating layers interconnection.

General procedure:

After drill; non interconnect between layers & need conductive layer on hole's wall. This process so called PTH (Plating Through Hole). There are three main processes contain: De-smear; Electro-less Cu deposition; Electrolytic Cu panel plating.

De-smear will remove smear & make micro rough on hole's wall in the same time. That can improve Cu with resin bonding. Electro-less Cu deposition will create thin metal on surface as the electrolytic plating seeds layer. Some makers applied other conducting media, like conductive polymer or graphite, are so called "Direct Plating Process". Electrolytical plating add more copper on the hole's wall. As at this point, there is non-pattern on boards, the step was called "Panel Plating".

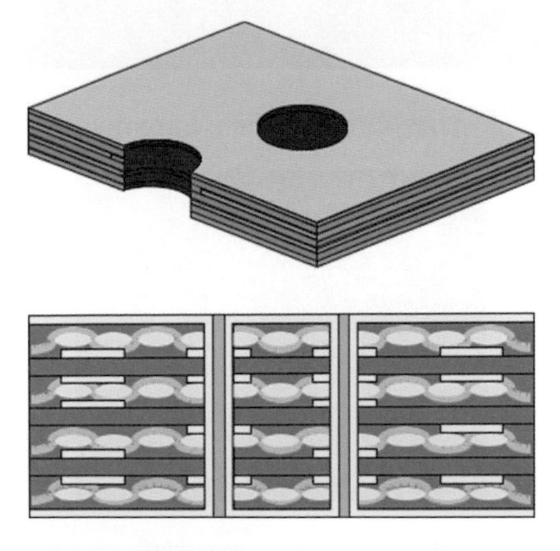

全板銅電鍍 / Panel Plating

外層影像膜貼合

製程說明

製作外部線路前的影像轉移膜製作。

工作內容

完成全板電鍍後，通孔內含有一定厚度的銅。在細線路製作方面，由於細線製作能力受限於未來蝕刻的銅厚度，愈薄的底銅能呈現愈好的細線蝕刻能力。因此許多業者，會用較薄全板電鍍來保證孔內導通性，再利用影像轉移及線路電鍍，在板面形成感光膜，做影像轉移及後續線路電鍍工作。

外層壓膜機 / Dry Film Laminator

壓膜完成板 / Board after Dry Film Laminator

Outer Layer Resist Lamination

Process description:

Create resist layer on surface for pattern plating.

General procedure:

After panel plating, through holes are deposited with base copper. Fine line capability will be limited by base copper thickness.

The thinner base-copper the better fine line etch capability. Many makers used low built panel plating & then apply plating resist for pattern plating as the outer pattern creation processes.

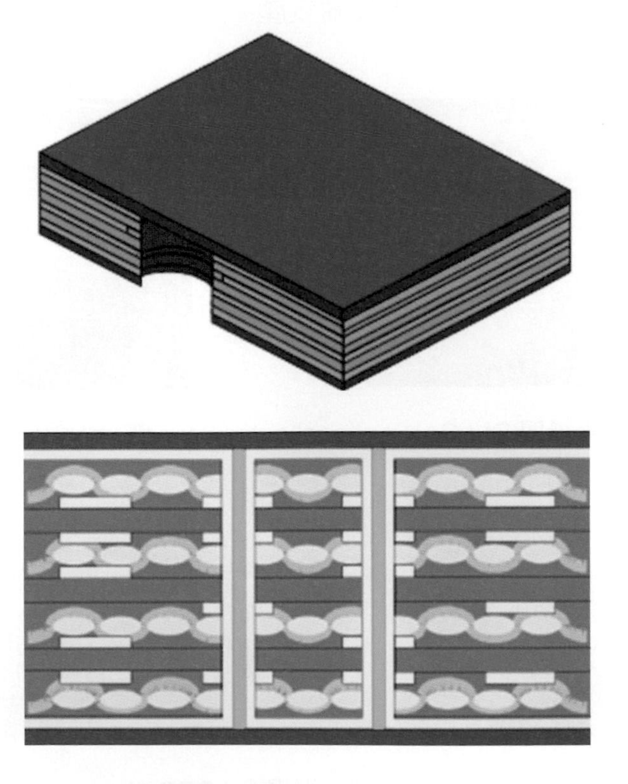

光阻乾膜壓合 / Dry Film Lamination

外層線路曝光

製程說明

外層影像轉移。

工作內容

感光膜貼附後，會經過類似內層板製程，再次曝光、顯影。但此次的光阻覆蓋區域與內層恰相反，因爲光阻主要功能是定義需要電鍍的區域，而覆蓋區域是不要電鍍的區域。所以這種光阻稱爲“電鍍光阻劑”。外層曝光機分兩段曝光，當然仍然可用雙面同時曝光，但爲了內部已有線路的對位，還是選用分面曝光爲多。目前高解析應用，已經以不少業者開始採用 DI 設備量產。

傳統外層曝光機與 DI / Conventional outer Layer Exposure Machine & DI

去膜、蝕刻、剝錫液 / SES Line

Outer Layer Exposure

Process description:

Outer layer image transfer.

General procedure:

Resist laminated, boards will repeat exposure; developing steps similar as inner layer. But coverage area will be different, resist open area will for plating. This was so called “Plating Resist”.

Most makers use side by side exposure in this step. Of course, maker can perform two-side expose in one shot. But without special care and design in machine, the registration won′t be that good as side by side. High registration application now applied more DI equipment for mass production.

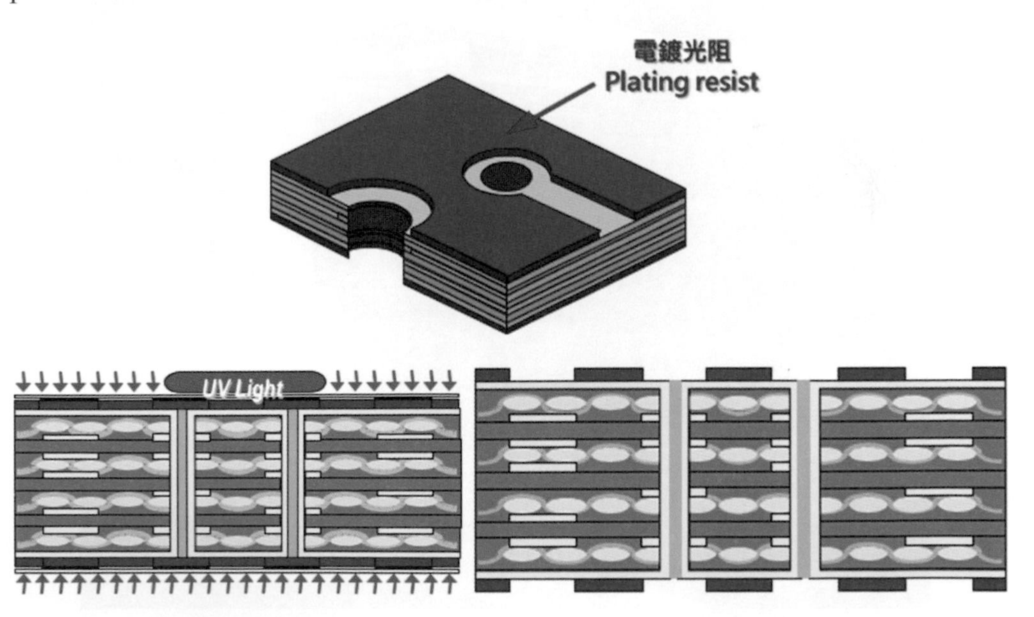

外層線路曝光
Outer Layer Exposure

外層線路顯影
Outer Layer Developing

線路電鍍

製程說明

線路及蝕阻電鍍。

工作內容

經過影像轉移的電路板，未來所需線路已曝露出來。此時將吊入線路電鍍設備，在線路區鍍上銅，此時孔內銅會繼續加厚，而孔銅加厚才是電鍍主要目的。因此在孔銅鍍足後，即停止銅電鍍，並同時鍍上一層錫，作爲後續蝕刻阻劑，這個整體過程稱爲“線路電鍍”。

線路電鍍線 / Pattern Plating Line

Pattern Plating

Process description:

Pattern plating & etching resist plating.

General procedure:

After outer layer image transfer, the pattern area was exposed. Boards will then be loaded in plating equipment for pattern and through holes plating.

The primary function of Cu plating is increasing copper thickness in holes, then after build enough copper in hole, the boards will be coated with thin layer of solder or tin as etching resist. The total process above was so called "Pattern Plating".

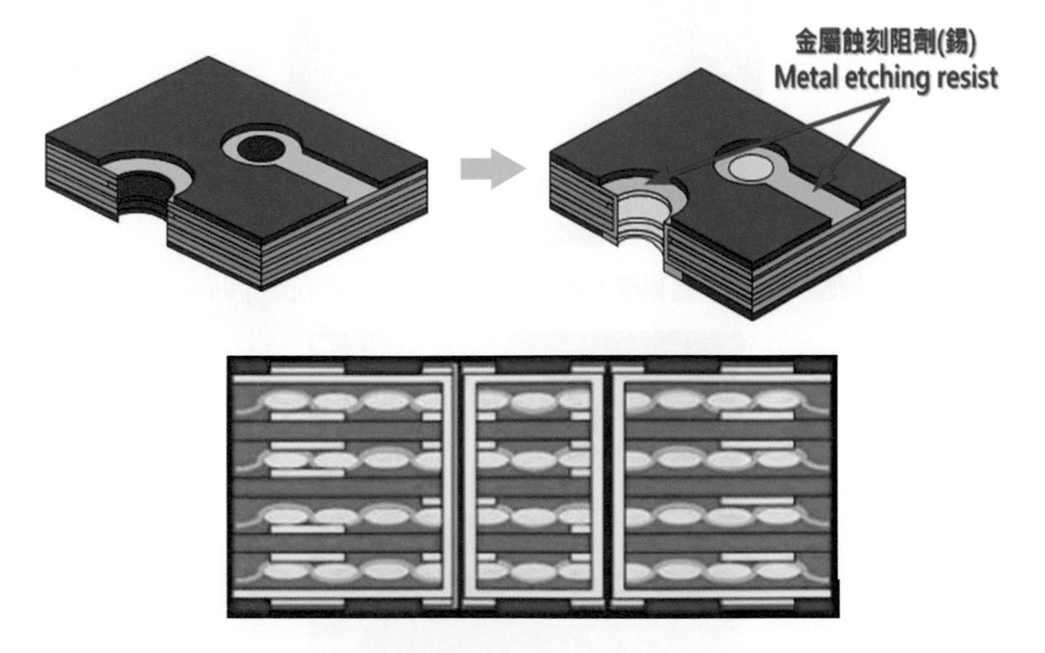

線路電鍍 / Pattern Plating

外部線路的形成

製程說明

製作外部線路。

工作內容

線路電鍍完成，送入剝膜、蝕刻、剝錫線，主要工作是將電鍍阻劑完全剝除，將要蝕刻的銅曝露在蝕刻液內。由於線路區頂部已被錫保護，線路區線路就能保留下來，如此整板的表面線路就呈現出來。圖面所示生產線就是典型剝膜、蝕刻生產線，業界統稱爲“SES Line”。

剝膜、蝕刻、剝錫線 / Stripping; Etching; Stripping Line

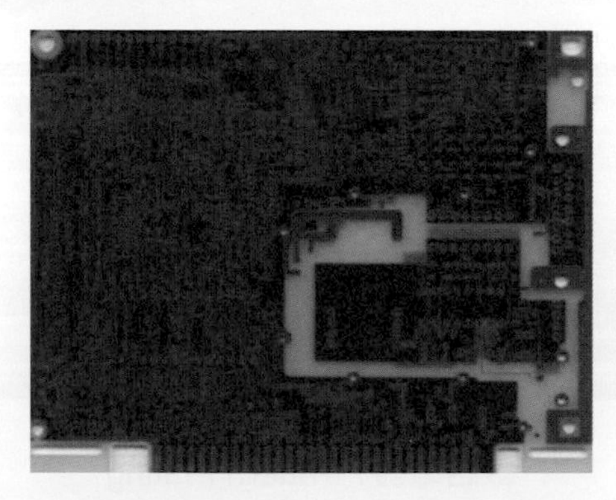

剝膜、蝕刻、剝錫後外層線路 / Post SES Outer Pattern

Outer Layer Pattern Creation

Process description:

Process for creating outer layer pattern.

General procedure:

After pattern plating, boards will get in film stripping; pattern etch; tin stripping.

Concepts are removing plating resist first and expose copper in etching solution. Then applied alkaline etching-solution as etcher. For tin can survive in solution for a while.

After pattern etching, tin resist will be removed and then leave the pattern on boards. The typical process are so called “SES Line”.

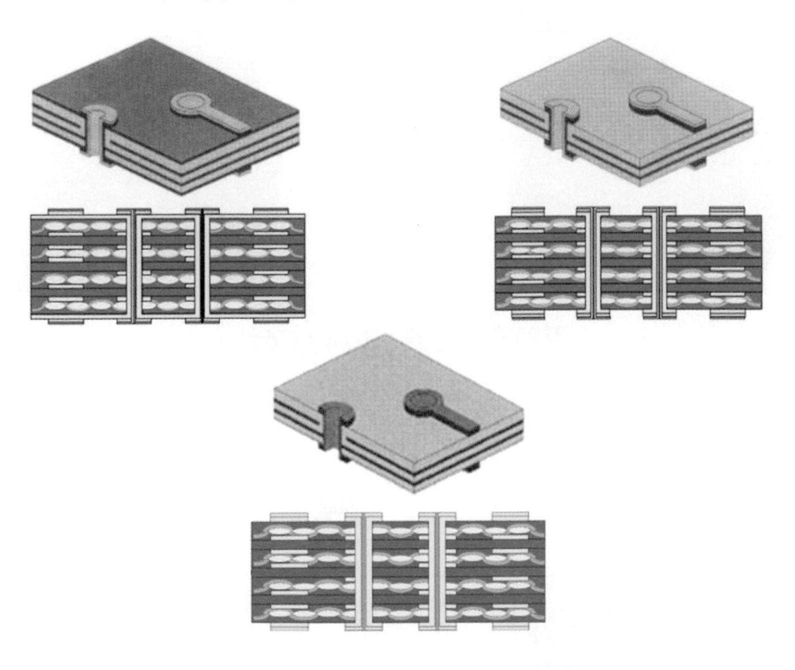

剝膜、外層線路鹼性蝕刻、剝錫

Film Stripping; Alkaline Etching; Tin Stripping

防焊漆塗佈

製程說明

防焊漆塗佈程序。

工作內容

電路板是爲承載、連接電子部件而作，線路完成後要將組裝區定義出來，其它部分以高分子材料保護。部件組裝多用焊錫，保護材料被稱爲“止焊漆”。早期以絲網印刷處理，但因部件小型、精準化，精密板必須以曝光製作。多數電路板，都已使用感光止焊漆。處理程序與乾膜製程略不同，爲了成本多數使用油墨。操作以印刷或全面塗佈，利用滾筒、簾幕式塗佈機完成塗裝。烘乾後，曝光顯影將保護與組裝區分開，圖面爲典型簾幕式塗佈機。

滾筒式塗佈機 / Roller Coater

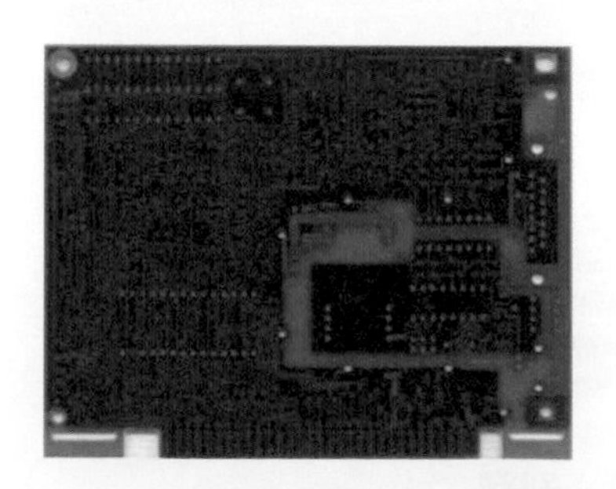

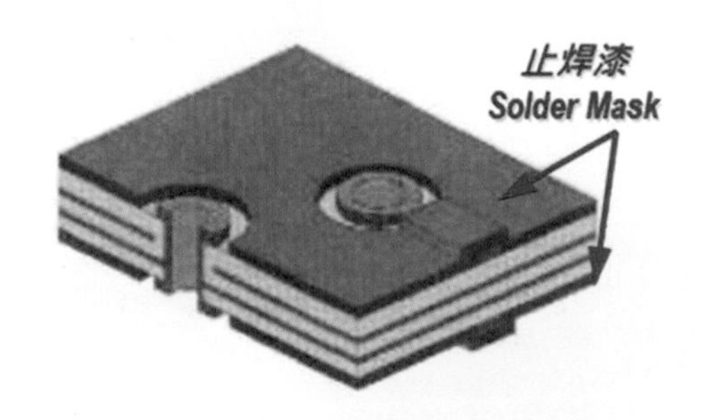

塗裝完成板 / Coated Boards

Solder Resist Coating

Process description:

Solder resist coating procedure.

General procedure:

PCB's main function was carrier & connecting parts. After patterning, non-assembly area has to be protected by material. That can protect the area & prevent damage or oxidation. These were partially coated polymer was so called "Solder Resist".

Early days, coating was made by screen printing. But screen-print have fine pattern & resolution issues. Now most of PCB are using photo sensitive solder mask for manufacturing. For cost issue, majority makers use liquid ink as resist. Coating by screen; rolling; curtain would be available. After de-solvent image transfer will followed & non-expose area could be developed. Picture below is typical curtain coater.

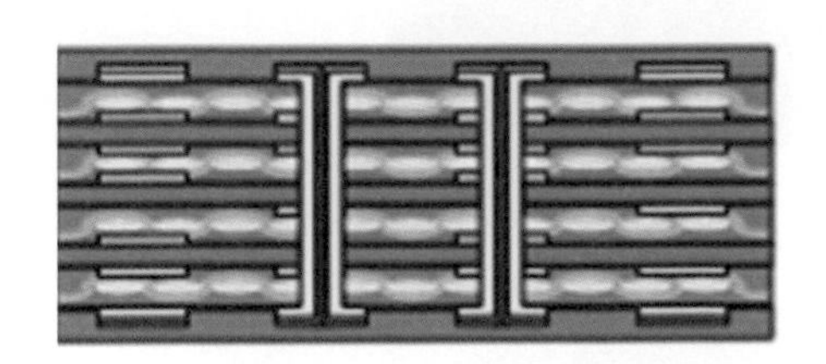

止焊漆塗佈 / SR Coating

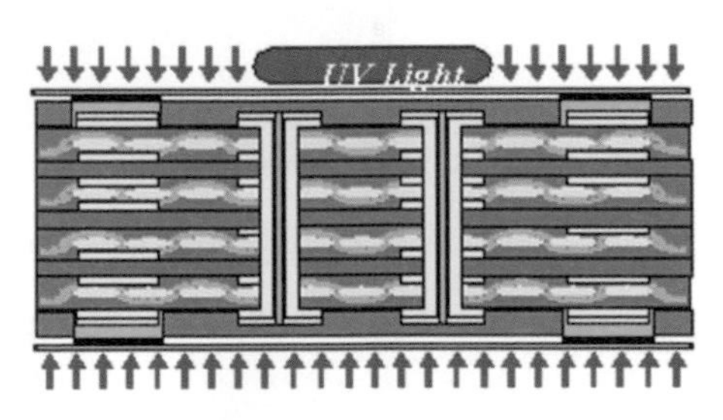

曝光 / Exposure

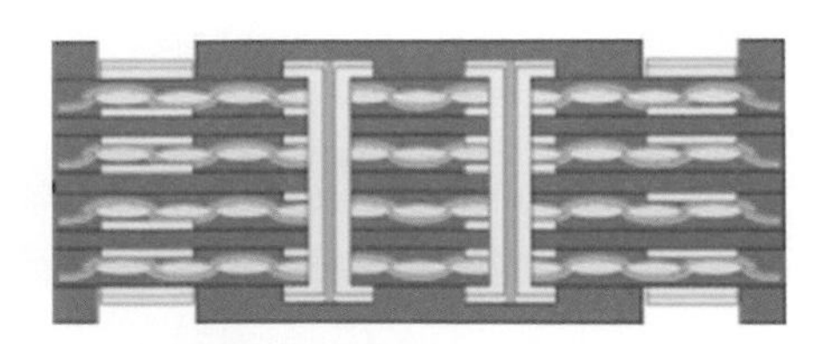

止焊漆顯影 / Developing

最終金屬化製程

製程說明

端子電鍍。

工作內容

電路板部分應用是在介面卡，因此金屬處理不僅止於焊接需求，端子部份仍須考慮插接組裝。因此電路板常在需要焊接區塗佈焊錫，塗佈方法以浸錫及熱風整平處理，這統稱爲噴錫 (HASL)。但介面卡端子區，會做金鎳金屬表面處理。下圖所示，爲典型端子電鍍設備與噴錫設備。

金手指電鍍機 / Gold Finger Plating Line

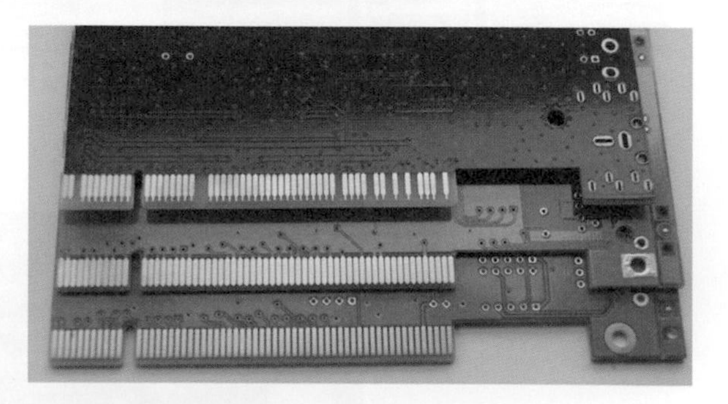

鍍金 / TAB Plating

Metal Finish(Hot Air solder Leveling & TAB Plating)

Process description:

Gold finger plating and other metal finish.

General procedure:

PCB interface card applications, soldering won't be the only one connection interface. Kinds of PCB need more than one surface finishing.

Solder dip PCB with air knife leveling process are so called "HASL (Hot Air Solder Leveling)" process. But the interface cards finger area need Nickel/Gold plating. Picture below is typical gold finger plating line & HASL equipment.

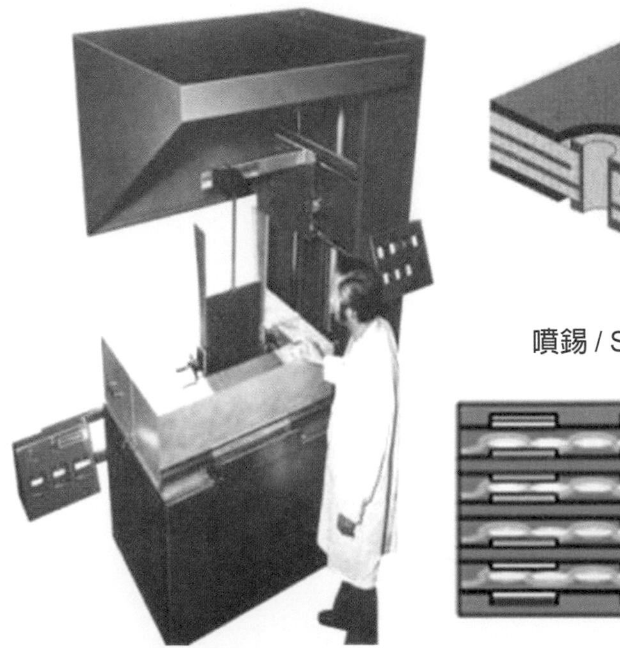

噴錫 / Solder Coating (HASL)

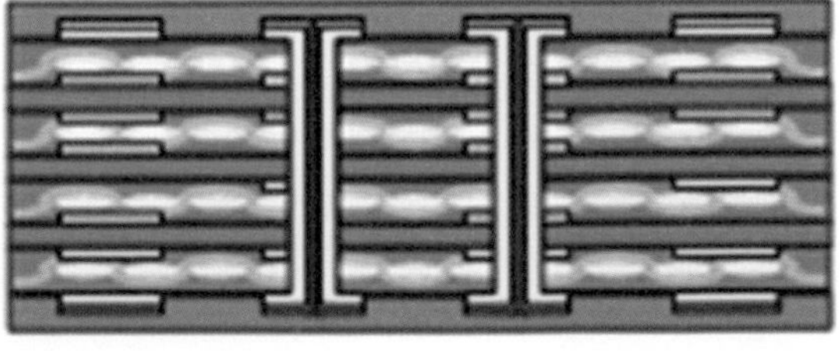

5

電路板基礎製程簡介－軟板篇

Basic Flexible PCB Manufacturing Introduction (Printed Circuit, FPC section)

軟板的應用

軟板應用，主要重點在材料柔軟可撓曲特性，以此進行產品製作。一般對軟板應用的考慮，包括：功能、成本、信賴度、構裝效率、空間應用等。良好的軟板設計可降低成本、改善功能、節省空間，這些影響都會被同時考慮。

FPC applications

FPC applications typically will use their flexibility to make the products.

Normally, the FPC applications will consider the Functions; Cost; Reliability; Packaging and Space using.

Good FPC design will have cost down; function improving and space saving benefits. These effective issues will be considered in the same time.

典型的軟板應用範例 / Typical FPC Application

軟性電路板的優勢

可撓曲性	穩定重覆撓曲性有助於特異形式電子組裝，產品的線路設計可以進行三維 (3-D) 的架構。
降伏性	由於軟板具有一定的降伏性，可以隨熱應力的變化自行調節接點間的距離，可以降低應力集中於接點的斷裂風險。
薄型特質	因爲材料薄而可以提供非常良好的撓曲性，同時因爲薄的材料使得熱傳效果良好，有利於機構的設計及產品的熱管理。
高溫的功能性表現	PI 在高溫下操作的表現良好，對於一些特殊的產品需要高溫操作時可以使用這樣的材料設計。
蝕刻與機械加工性	部分軟板材料可用雷射或化學蝕刻進行開窗作業，這對於單層銅皮兩面組裝的產品言十分重要。
互連的可能性與尺寸縮小重量減輕	軟板可以撓曲依據需要進行立體彎折，採用軟硬板製作方式又可以減少端子雜訊及信賴度的問題，因此可以簡化連結省下連結器及端子，因此產品重量也會下降。
結構簡化空間利用率高	軟板可替代許多的點對點連線元件，將不同平面無法連結的線路連結，因設計可以簡化空間的利用率得以提高。

FPC Strengths

Excellent Flexibility	Stable and repeating flexibility is helpful for 3-D electronic products assembly.
Lower Toughness	Lower toughness is benefit for standing more thermal stress and reducing the joint cracking risk.
Thin dielectric	Thin dielectric could have better flexibility and better heat transfer, those will be benefit for structure design and thermal management.
Great high temperature performance	PI material could be operated under high temperature environment then can fit for high temperature applications.
Good manufacturability	Some flexible material can use laser or chemical etch create pads or window then could be used in double access applications.
Multi-plane connection and size shrinking	Flexible boards could be bended and binding with rigid-flex technology then can reduce connector and terminator using that can reduce weight.
High space utilization	Flexible boards can replace many connecting parts and join multi plane boards then by these benefits can saving space and make design easier.

一般軟板製作流程與產品

Typical FPC Boards Manufacturing Processes & Production sequences.

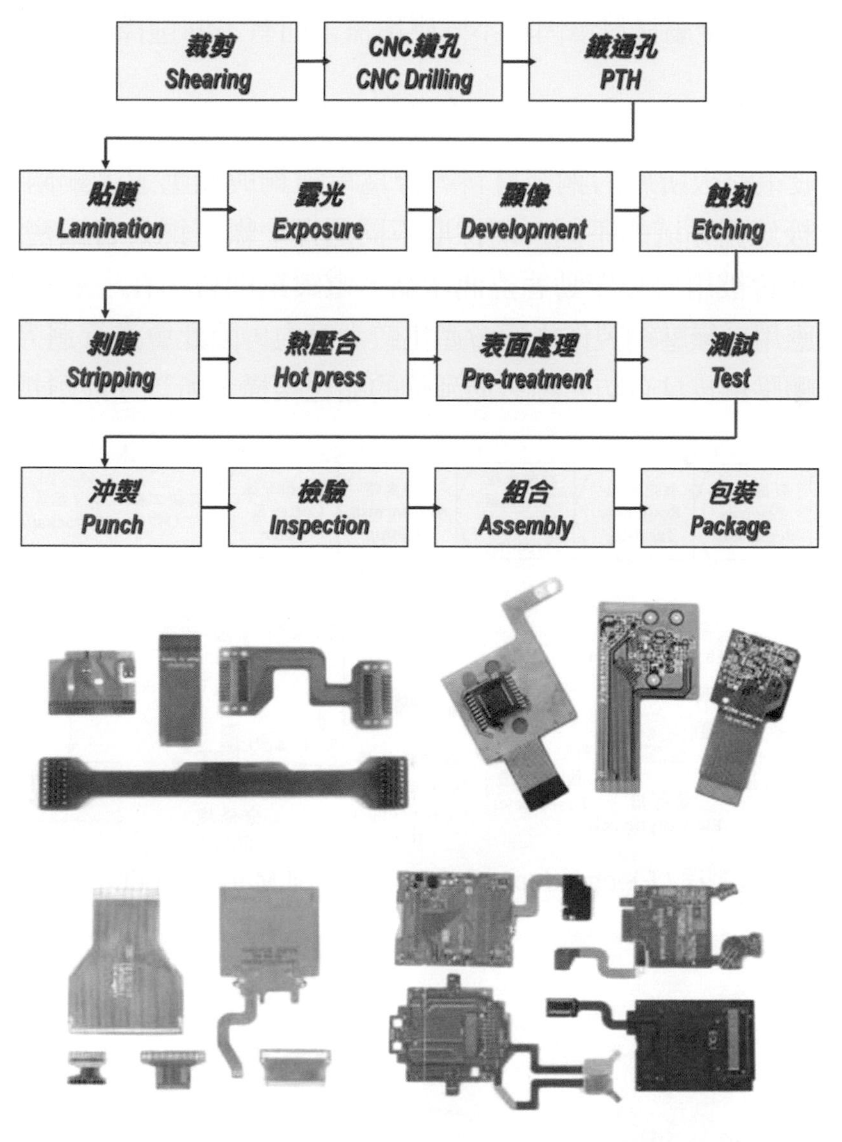

典型的軟板產品 / Typical FPC Products

軟板材料所用的銅金屬材料差異

製程說明

軟板中銅金屬材料會因爲不同應用需求而有不同選擇。

工作內容

一般電路板所用的銅箔材料，分爲電鍍銅與輾壓迴火銅兩種。會因爲軟板應用領域的不同，而採取不同選擇。輾壓銅具有緻密耐折特性，因此會被用於製作動態撓曲產品，電鍍銅則會用在一般組裝非動態撓曲應用。輾壓銅製作過程會產生較大應力，因此會進行迴火處理，迴火後輾壓銅皮具有防止斷裂面延伸的晶粒結構，所以才能耐折。

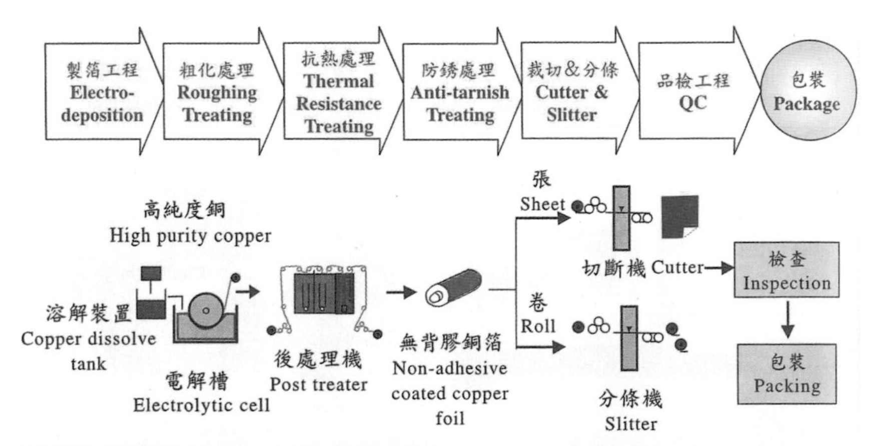

電鍍銅皮製程 / Electro-deposited Copper Foil Manufacturing Process

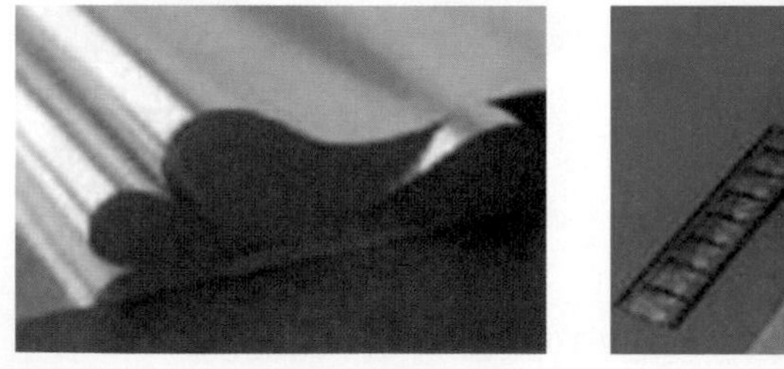

電鍍銅皮與輾壓銅皮 / Electro-deposited Copper Foil & Rolling Foil

Different copper foil used in FPC

Process description:

FPC used copper foil, the material selection will depend on applications.

General procedure:

Normal PCB used copper foil material could be category as electro-deposited and rolling two types. As the different application maker will chose them separately.

Rolling copper foil is tougher and then can be used in dynamic applications. Electro-deposited foil could be used in non-dynamic applications. Rolling copper foil manufacturing process will create much stress inside and then need annealing. By annealing, RA foil has suitable grain structure to prevent crack propagation.

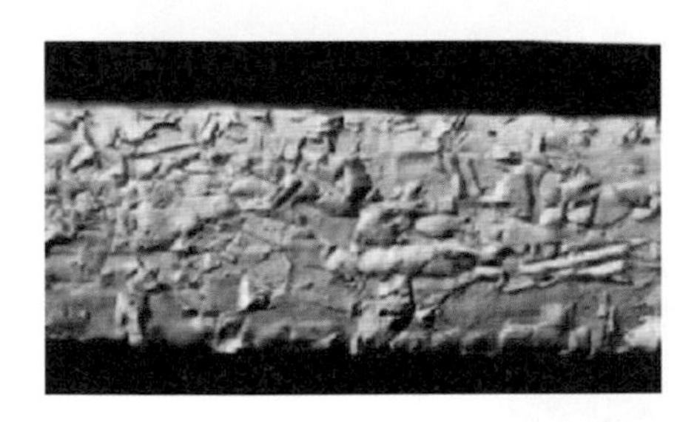

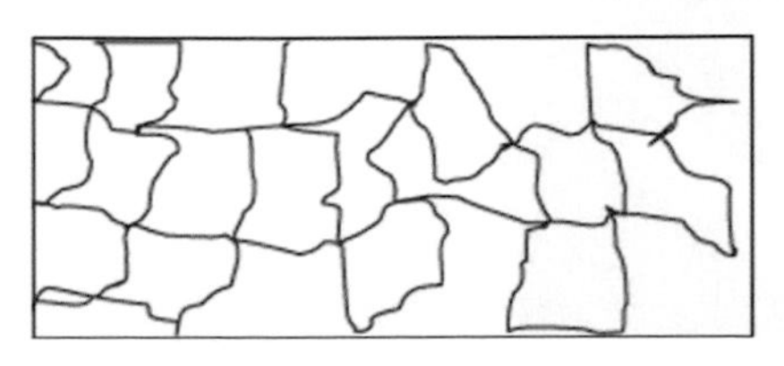

輾壓銅皮 / Rolling Foil

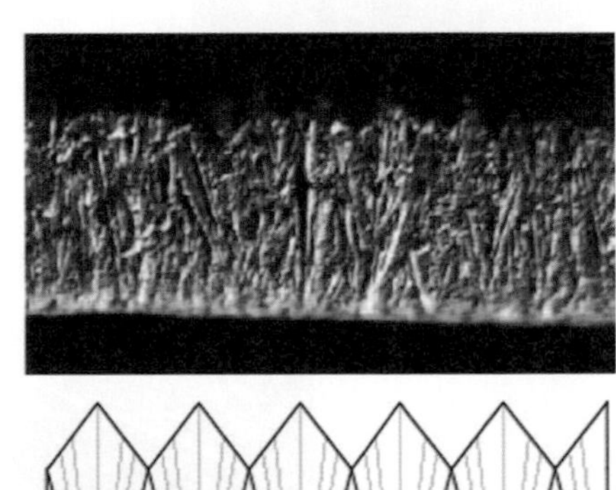

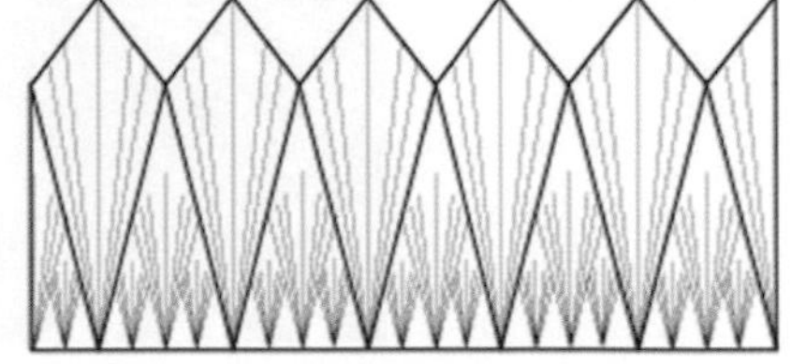

電鍍銅皮 / Electro-deposited Copper Foil

材料裁切

製程說明

軟板材料的前置處理。

工作內容

一般軟性電路板材料，都以捲狀製造。爲配合不同尺寸需求，必須搭配規劃切割出最佳利用率尺寸。依據材料配置規劃，製作軟板第一步驟就是將捲狀材料切割爲單片材料進行生產。也有些大量生產軟板產品，採用捲對捲生產，此時就不需要進行單片切割。

裁剪 / Slicing Machine

Film Slicing

Process description:

Raw material pre-cutting.

General procedure:

Most of flexible circuits raw materials are roll format normally.

For different demanding, users have to optimize the utilization. The first step is slicing film to working size.

Some products will be applied with reel -to-reel manufacturing & then the slicing procedure will be eliminated.

聚亞醯胺樹脂薄膜 / PI Film

銅皮/Copper Foil
接著劑/Adhesive
基材/Base Material

單面板銅箔材料結構

Single Sided Material Structure

銅皮/Copper Foil
接著劑/Adhesive
基材/Base Material
接著劑/Adhesive
銅皮/Copper Foil

雙面板銅箔材料結構

Double Sided Material Structure

機械鑽孔

製程說明

產生層間導通孔。

工作內容

一般雙面或是多層的軟性印刷電路板，會依據客戶的需求進行層間的導通加工。在軟板的製作上，會進行定位孔、測試孔、零件孔等等的製作，之後延續到下一個製作程序。

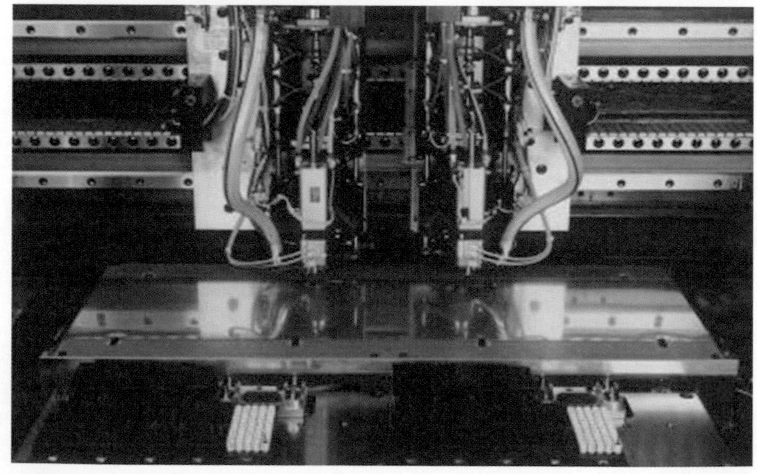

軟板捲對捲鑽孔機 / FPC Reel-to-Reel Drilling Machine

CNC Drilling

Process description:

Creating via between layers.

General procedure:

Normal double sided and multi-layer flexible boards will follow the customer′s inquiry to create layer connecting via.

In flexible boards manufacturing will make like alignment hole; testing hole; components hole and so on and then pass through the post processes.

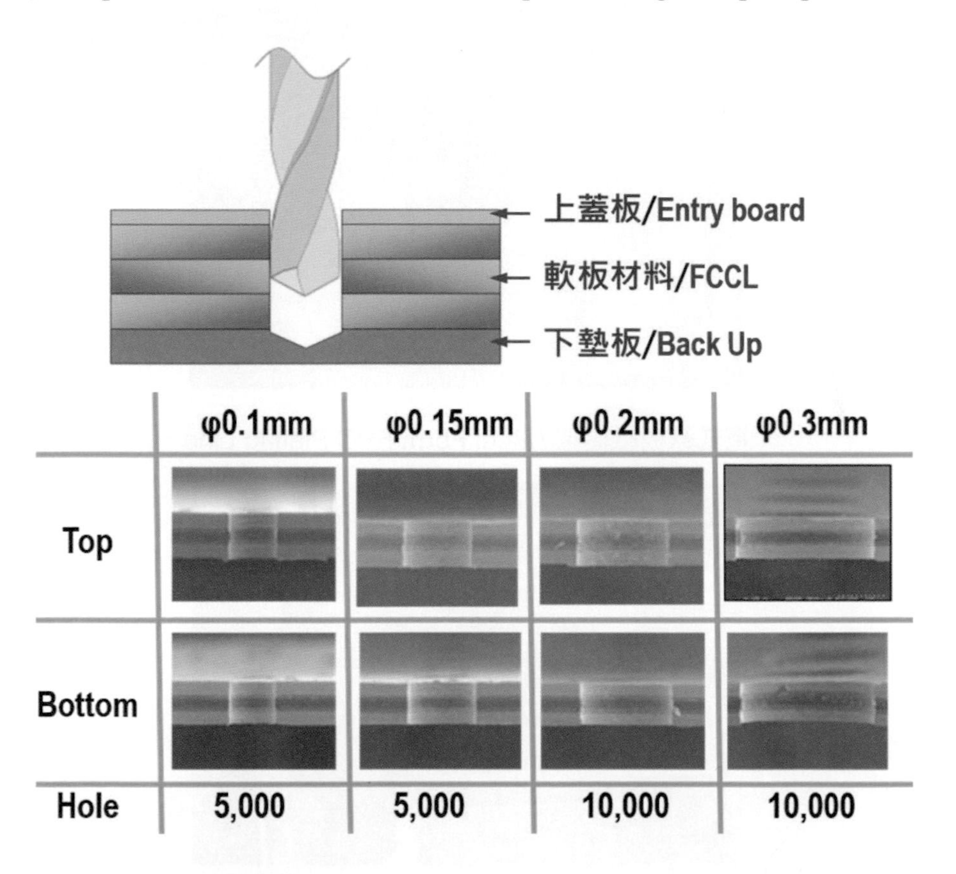

	φ0.1mm	φ0.15mm	φ0.2mm	φ0.3mm
Top				
Bottom				
Hole	5,000	5,000	10,000	10,000

機械鑽孔成果 / Mechanical Drill Performance

鍍通孔

製程說明

進行導通孔的金屬化。

工作內容

鑽孔完成後，上下層間的導體並未導通，因此須在孔內的孔壁上形成一層導電層以導通訊號。鍍通孔是有雙面導體以上的產品才有的製程。

捲式軟板電鍍線 / Reel Form FPC Plating Line

片式軟板電鍍線 / Panel Form FPC Plating Line

Plating Through-Hole

Process description:

Metallization through holes.

General procedure:

After drilling; between layers are not connected yet. Makers have to create a conductive layer for single transformation.

The functions of Plating-through-hole processes are just for double sided plus flexible boards making.

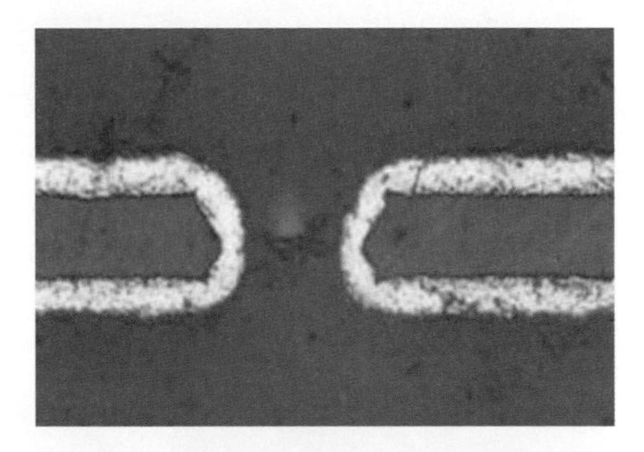

雙面軟板通孔電鍍成果 / Double Sided Flexible Boards Plating

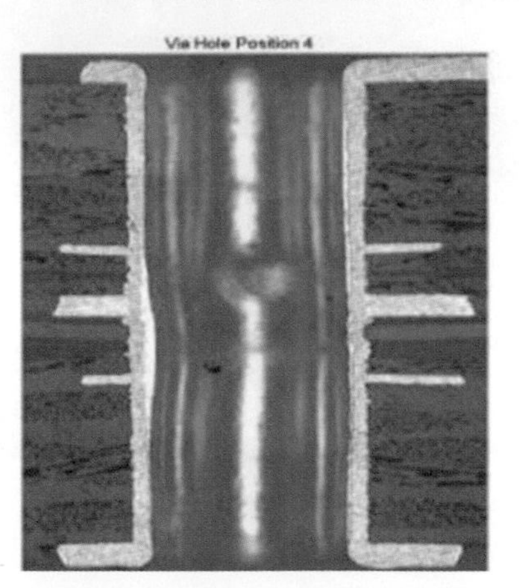

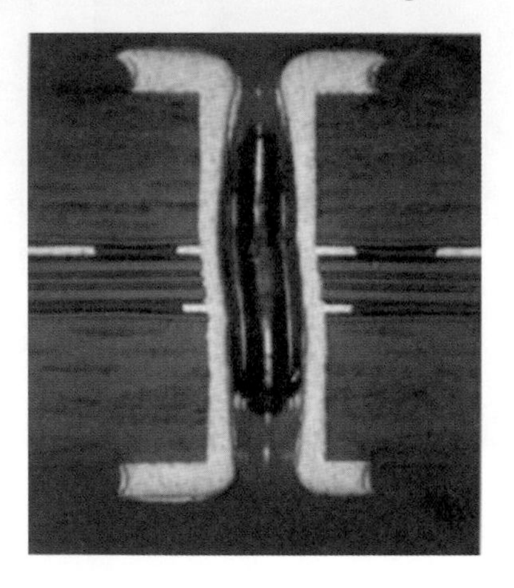

軟硬板通孔電鍍成果 / Rigid Flexible Boards Plating

貼膜

製程說明

感光膜貼附。

工作內容

鍍通孔完成後，以熱滾輪加壓貼合的方式，在已經清潔完成的材料表面貼合一層乾膜 (Dry Film) 作為蝕刻阻劑。由於乾膜對光線中的紫外線敏感，因此必須在黃光室中作業，以避免乾膜在製作前就發生變化。

軟性電路板在大量生產的模式中，常會使用捲對捲的生產的方式，這樣的操作，貼膜作業所使用的設備就與一般單片操作的設備略有不同。

捲式連續壓膜機
Reel to Reel Laminator

手動壓膜機
Manual Laminator

Dry Film Lamination

Process description:

Photo film lamination.

General procedure:

After PTH boards will pass through hot roll dry film lamination. Dry film will be laminated on pretreated boards as the etching resist. Dry-film are kinds of photosensitive film so the operation has to proceeding in a yellow light clean room for preventing side reaction.

FPC mass production might use reel-to-reel mode. This film laminating operation is different from panel to panel mode and need different laminating equipment.

捲式連續壓膜設備連線 / Reel to Reel Laminating & Treating Line

曝光

製程說明

對感光膜進行影像轉移的感光處理。

工作內容

貼膜完成的材料，上面的線路是使用影像轉移的方式，將設計完成的工作底片上的線路型式，用紫外線曝光的方式，將線路型式移轉到乾膜上。如此一來，在曝光完成的乾膜上，就會留下感光膜線路外型。軟板的工作底片是以負片製作，透明的區域就是線路及留銅部分。

手動對片 / Manual Mylar Alignment

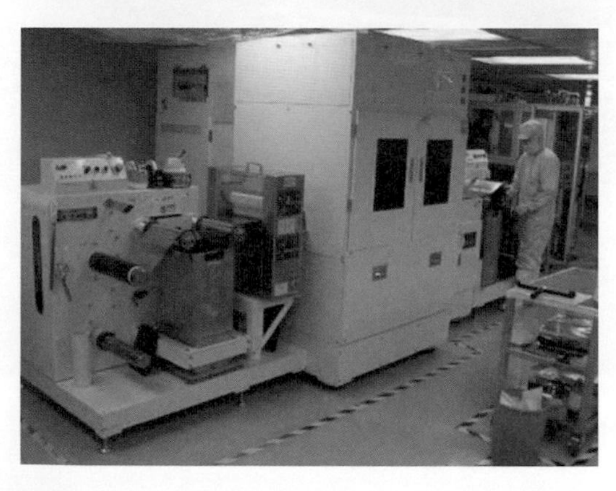

捲對捲自動曝光作業 / Reel to Reel Auto Exposure Operation

Exposure

Process description:

Photo film passed through exposure step.

General procedure:

Laminated boards will loaded on image transfer process. Through UV exposure process the image will transfer from artwork to the photo film. Flexible PCB is manufacturing with negative film that the light passing area will leave on boards after developing process. The area will be copper pattern area.

板面清潔機
Boards Surface Cleaning Machine

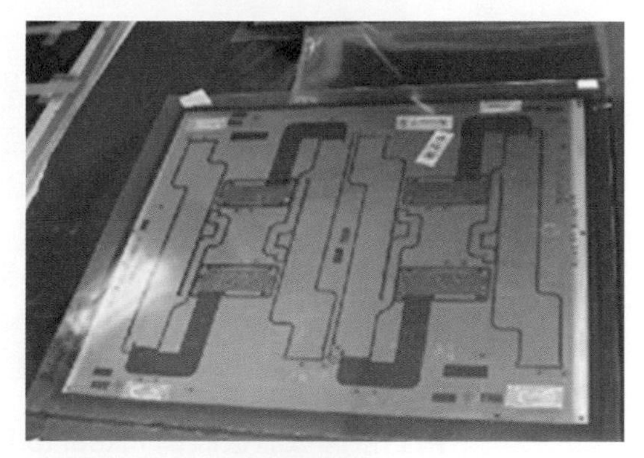

手動曝光
Manual Exposure

捲對捲曝光機構 / Reel to Reel Exposure Mechanism

顯影

製程說明

將未見光的感光膜去除。

工作內容

經過曝光的乾膜會聚合硬化，未曝光部分則否。使用特定的藥水沖洗，將會把未經曝光硬化的乾膜沖洗掉，露出材料上的銅層，讓蝕刻藥水能接觸銅面。在此階段完成後就可以在材料上看出線路的形狀。

捲對捲顯像放板設備 / Reel to Reel Developing Loader

軟板顯影 / FPC Developing

Development

Process description:

Remove non-exposure photo film.

General procedure:

Through exposure the light absorbing area′s polymer will cure. But non-exposure area can be removed from boards and copper will expose on those area. Applying etch solution; these areas will be etched off. We can distinguish the pattern after developing.

捲對捲線路顯影 / Reel to Reel Imaging Pattern

捲對捲顯影設備 / Reel to Reel Developing Line

蝕刻

製程說明

將未被覆蓋的區域銅金屬蝕除。

工作內容

曝光完成的基板經過蝕刻藥水沖洗，會將沒有乾膜保護的部分銅層蝕刻去除，而留下有硬化乾膜保護的線路。

典型捲對捲軟板線設置 / Typical Reel to Reel Wet Process Setting

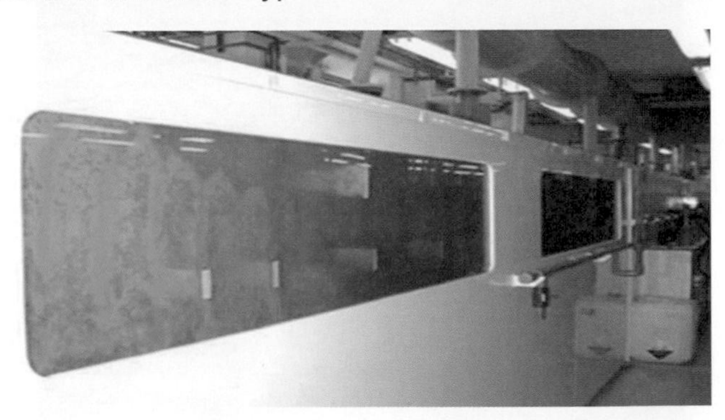

軟板蝕刻段 / FPC Etching

Pattern Etch

Process description:

Etch off non-cover area copper.

General procedure:

By etching resist selecting then apply corrosion agent so call "Etching Solution" that non-protected copper will be moved away by chemical reaction. Those protected area will leave there.

Chemical Reaction / 化學反應

$$Cu + CuCl_2 \longrightarrow CU_2Cl_2$$

Copper Etching Reaction / 蝕銅反應

$$2Cu_2Cl_2 + 4HCl + O_2 \rightleftharpoons 4CuCl_2 + 2H_2O$$

Regenertion Reaction / 再生反應

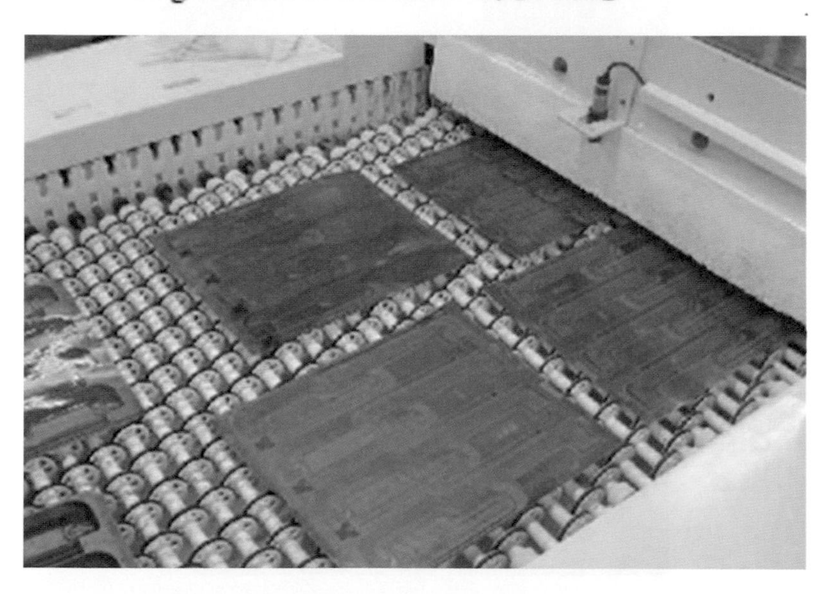

片式軟板線路蝕刻 / Panel Form FPC Pattern Etch

剝膜

製程說明

清除線路表面的乾膜。

工作內容

蝕刻完成的線路上仍然留有硬化的乾膜，在此流程使用特定藥水沖洗基板，利用藥水的特性清除附著在線路上的乾膜，露出完整的銅層線路。

捲對捲剝膜 / Reel-to-Reel Stripping

片式剝膜 / Panel Form Stripping

Film Stripping

Process description:

Trip off cured film on pattern.

General procedure:

After pattern etching the etching resist still on the pattern. Maker has to remove film residue and then pattern will expose after.

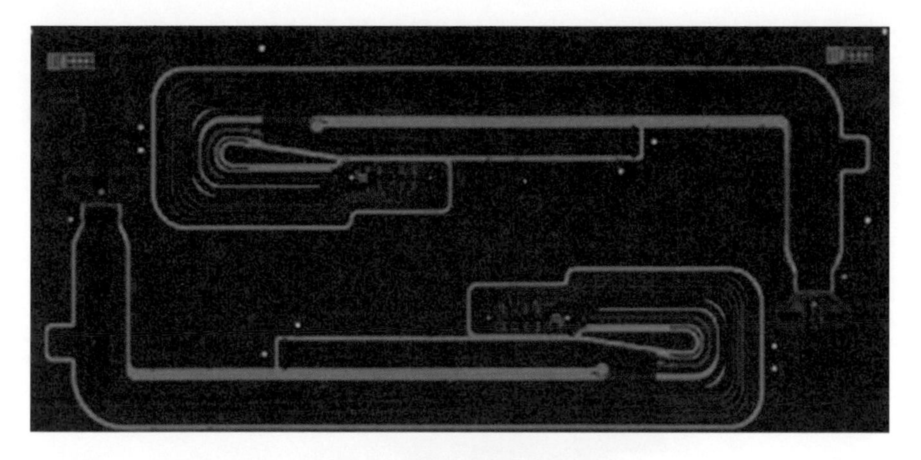

剝膜前 / Before Stripping

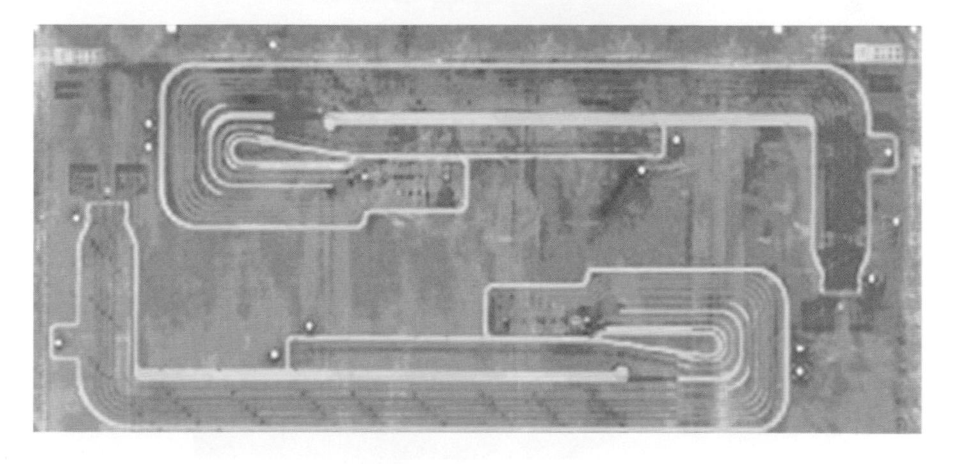

剝膜後 / After Stripping

假貼

製程說明

預黏覆蓋膜在軟板線路上。

工作內容

爲保護線路及客戶的需求，必須在導體上製作絕緣層，一般軟板使用的絕緣層稱爲”絕緣覆蓋膜 (Cover Lay)”。此流程的內容是將加工後的覆蓋膜對準位置，預先貼附在需要覆蓋的線路區域。

假貼 / Pre-lamination

Pre-lamination

Process description:

Pre-laminating cover layer on FPC pattern.

General procedure:

For protecting pattern or meet customer demanding then flexible boards will cover with insulating layer on pattern. The insulating layer is so called "Cover Lay". According to the operation are after material been cut to laminating size. The step of operation has to take care about the alignment.

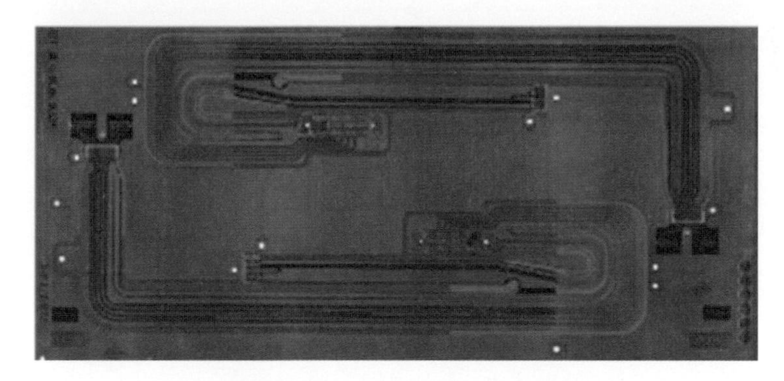

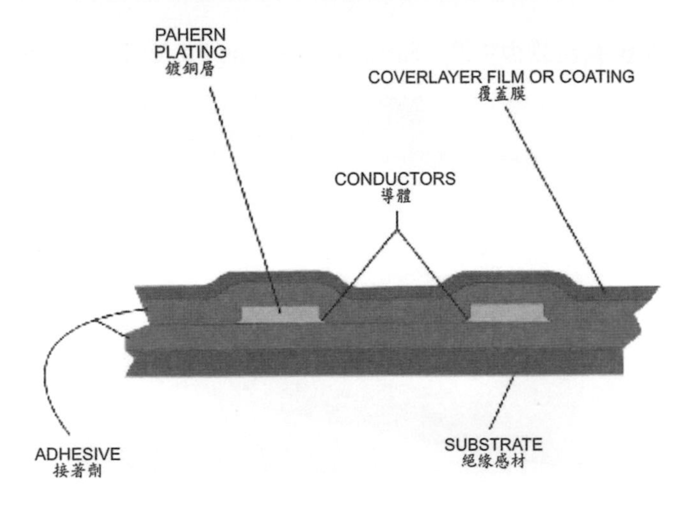

熱壓後的成品 / Board After Hot Lamination

熱壓合

製程說明

利用熱與壓力將貼合膠完全密合。

工作內容

絕緣覆蓋膜結構包含一層樹脂及一層絕緣膜。熱壓合則是利用機械提供熱能及壓力，將熔化的樹脂填滿線路間的縫隙，同時將覆蓋膜與基材緊密的貼合，形成完整的線路板。

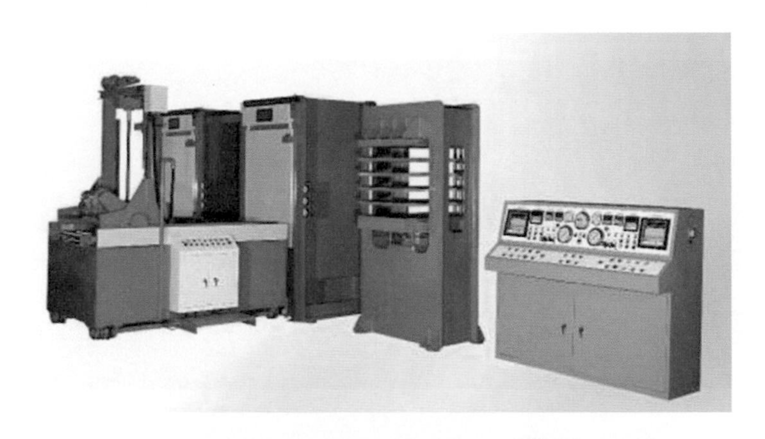

小壓機更適合軟板生產 / Small Hot Press Fit for FPC Making

一般快壓機 / Normal Quick Laminator

Hot Press Lamination

Process description:

Apply with heat and pressure to make bonding.

General procedure:

Cover lay′s structure including a layer of resin and a layer of insulator. So hot press will apply heat and pressure to melt resin and fill the spaces in between. After cover-lay binding tight then flexible boards operation has been complete.

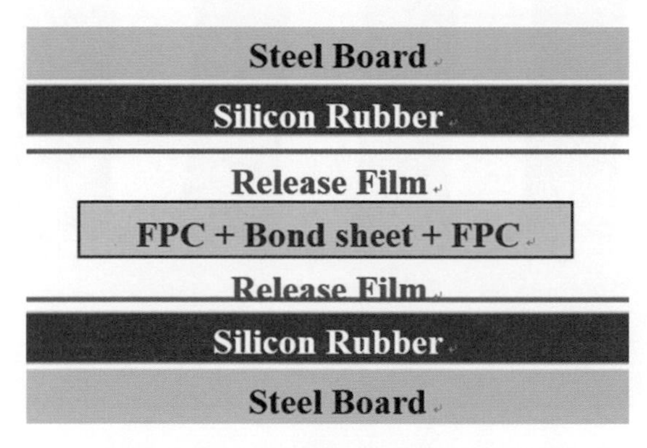

快壓堆疊結構 / Stacking Structure

真空快速壓機 / Vacuum Quick Laminator

表面處理

製程說明

爲軟板後續組裝所做的金屬表面處理。

工作內容

電路板的裸露位置，必須根據客戶需要，以電鍍方式鍍上鎳金或是錫鉛等不同金屬，以保護端子及確保線路性能。

不平整的端子覆蓋狀態 / Connector Edge Coverage is not Even

為熱壓式組裝所做的金屬表面處理 / Metal Finish for Hot Bar Assembly

Metal Finish

Process description:

Metal coating for assembly functions.

General procedure:

Bonding area will depend on customer′s inquiry to do metal finish. Those areas will be coated with solder or nickel and gold for wire protection and keep terminator′s function like solder-ability.

最終金屬表面處理線 / Metal Finish Plating Line

硬板與軟板間需要有良好的最終金屬表面處理

Rigid Boards & Flex Boards Interface Needs Good Metal Finish

電氣測試

製程說明

完成品的電氣測試。

工作內容

以探針治具及測試機台測量每片電路板是否有短斷路等不良情況。

針盤式測試機 / Nail Bed Tester

飛針測試機 / Flying Probe Tester

Electrical Testing

Process description:

Final products electrical test.

General procedure:

With probing testing machine detecting if the boards have any open/short defects.

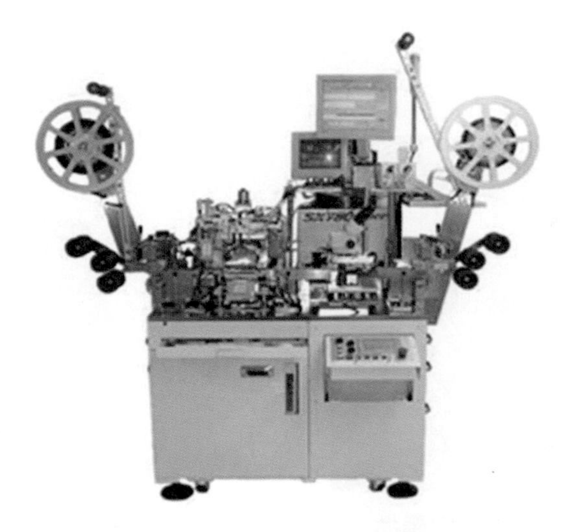

捲式非接觸自動測試機 / R to R Non Contact Auto Tester

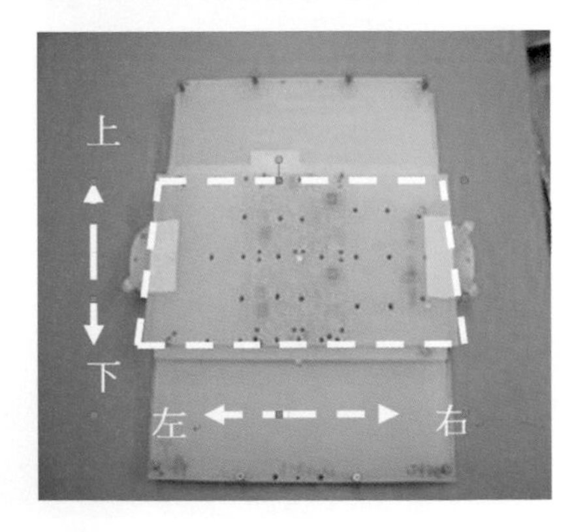

可微調的電氣測試治具 / Micro Adjustable Testing Tool

成形

製程說明

利用沖壓模具完成軟板外型製作。

工作內容

線路測試完成後，會利用鋼模或是雷射切割外型，將不需要的廢料及不要的線路分開。先進的雷射切割設備，已經有實力可以逐漸與低價的傳統切割競爭。

沖模有高準度及長使用壽命

Die have better accuracy & longer life

低結碳雷射切割機

Low carbonize laser cutter

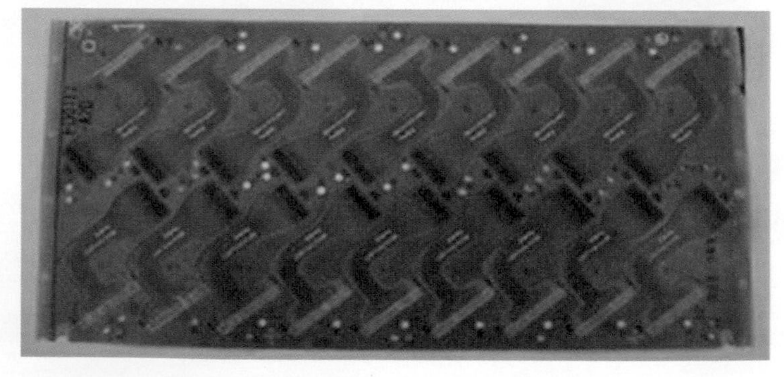

經過切割的軟板 / FPC after cutting

Contouring

Process description:

Applied with punch tool for FPC final sizing.

General procedure:

After testing, maker will use punching tool or laser trimmer creating the needed final size.

刀模 / Ruller Cutter

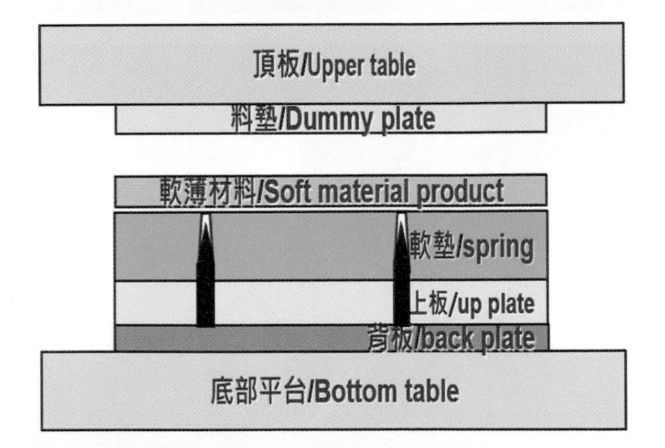

刀模成本低、精準度略差、壽命較短

Ruler Cutter have lower cost but shorter life and less accuracy

檢驗

製程說明

檢出不良品。

工作內容

檢查成形後的電路板，將線路內部有缺點但不影響導通性能及外觀不良的產品篩選出來。

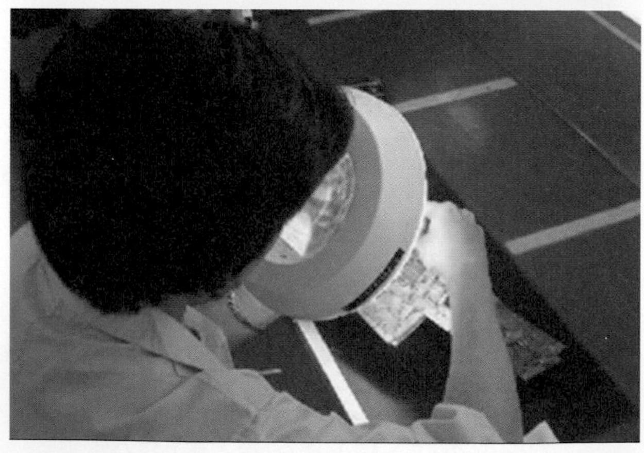

目視檢查 / Visual Inspection

Inspection

Process description:

Screening defect boards.

General procedure:

Inspect the boards after to filtrating the defects and scraped boards.

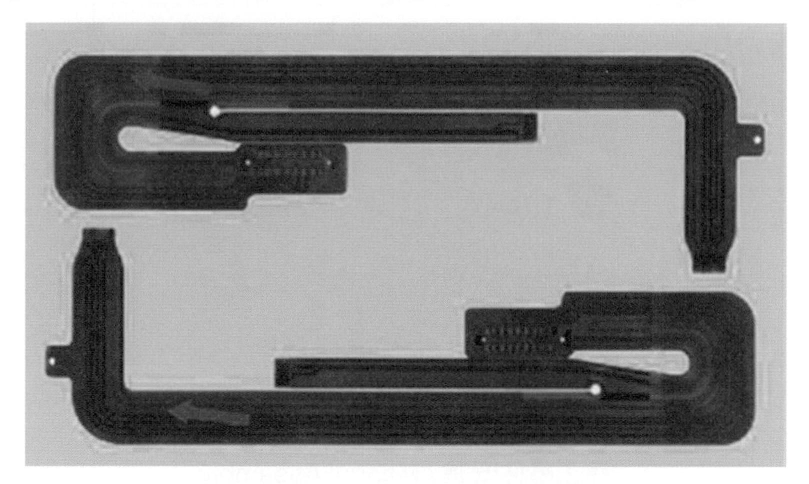

空板檢查 / Bare Board Inspection

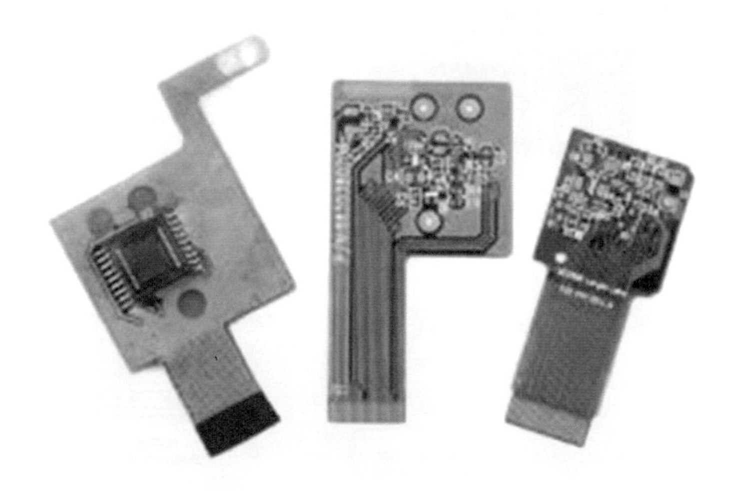

組裝後檢查 / Inspection After Assembly

組裝

製程說明

軟板完成後需要依據客戶的需求進行一些零件組裝。

工作內容

電路板上爲配合客戶組裝的需要，必須組合上各項零件，例如：背膠、遮光片、補強板等配件。

手工零件安裝 / Manual Assembly

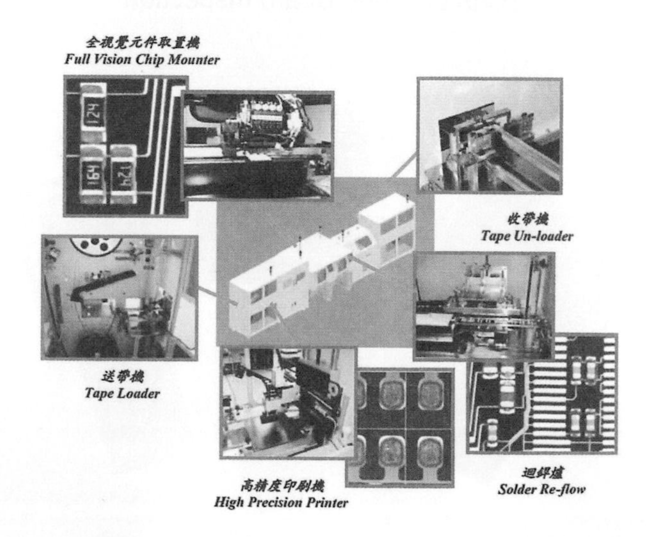

捲帶式自動組裝 / Reel to Reel Assembly

Assembly

Process description:

FPC has to follow customer′s demand to do some assembly.

General procedure:

By customer inquire assemble the parts and device by customer′s design. For example, like coating the backside glue/optical filter/back up plate and so on.

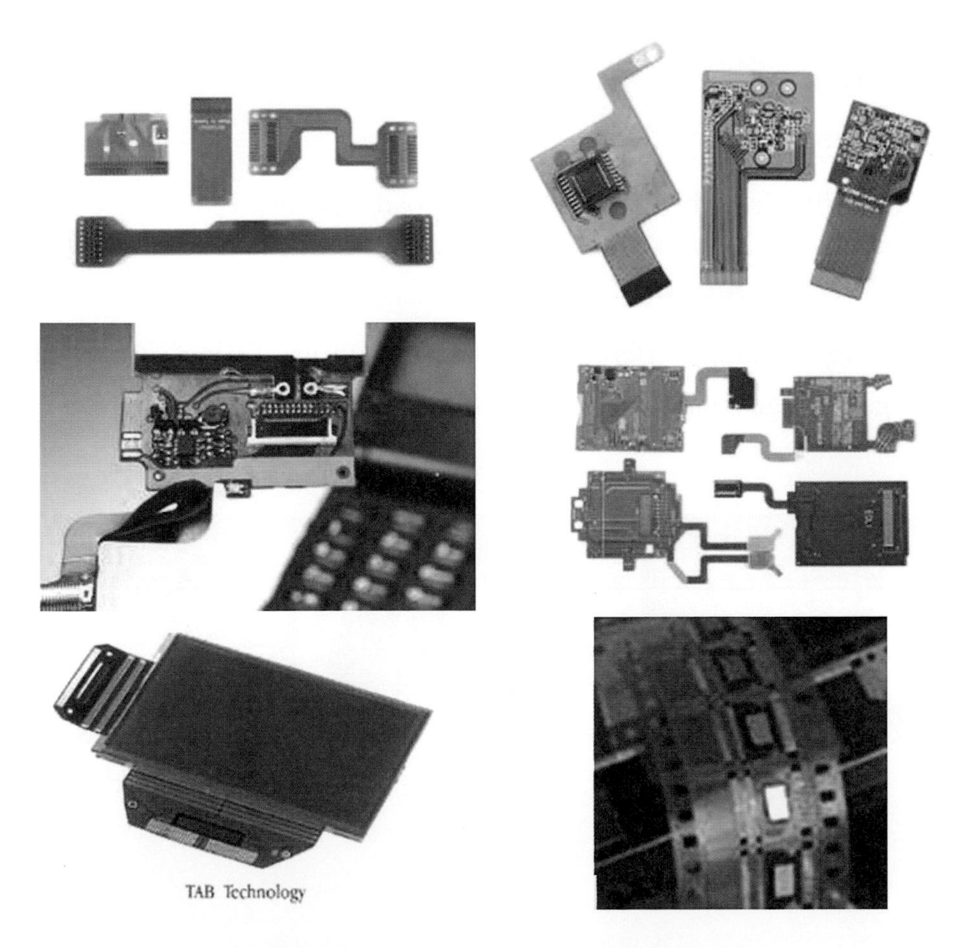

各式各樣的軟板組裝 / Kinds of FPC Assembly

包裝

製程說明

組裝完的軟板安放在固定包裝盒內。

工作內容

軟板出貨會依不同客戶需要及電路板形式，製作不同包裝盒來包裝電路板，以避免運送中造成損失。

Packing

Process description:

After FPC assembly, products will be loaded in tray.

General procedure:

Before shipping, different products have to use fitting case to pack. Packing has to guarantee no damage issue while transportation.

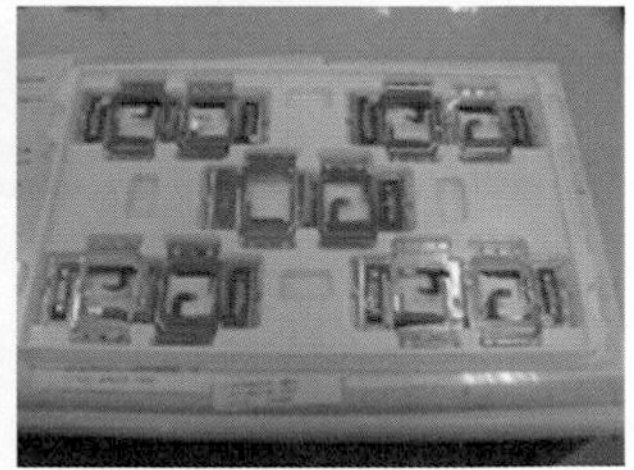

6

電路板的未來發展趨勢

PCB Future Trend

電路板業的產業定位

PCB Industry Role

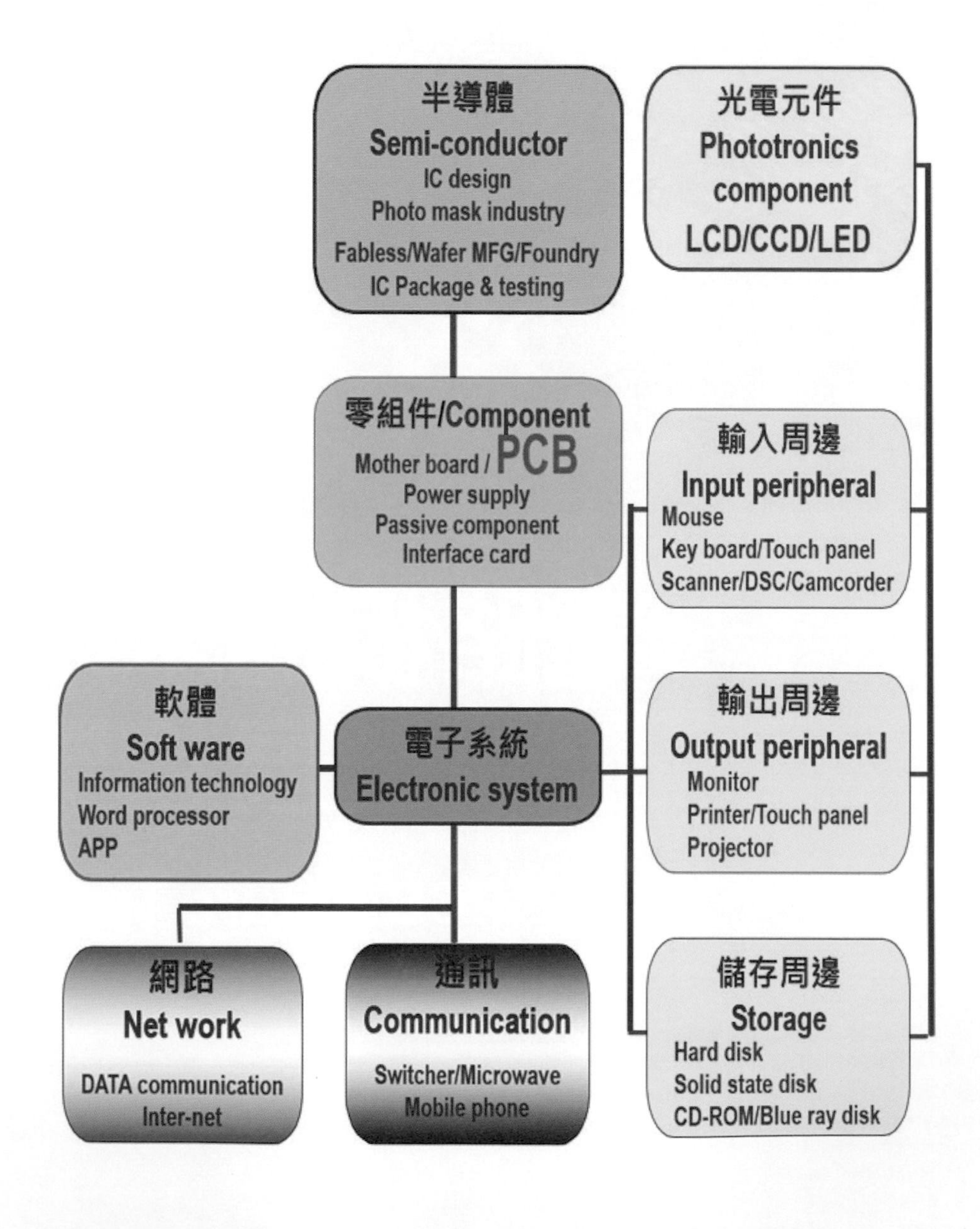

行動裝置

Mobile devices (Miniaturize for portable and wearable)

圖面所示爲典型的通信話機小型化、可穿戴化。

過去的小型化 / The mobile before

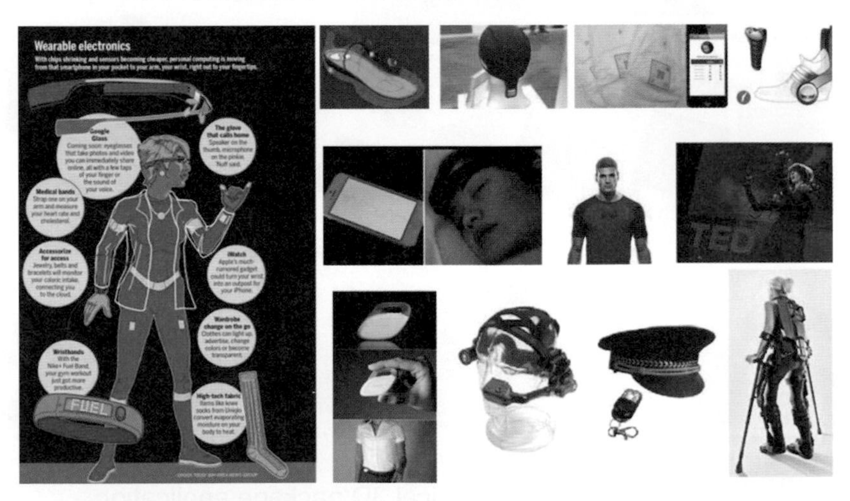

現在的可穿戴與未來… / The wearable now & the future…

電子元件構裝趨勢

電子產品輕薄短小化，重點總在電子部件構裝形式的微型化與尺寸壓縮。過去談導線架轉換到構裝載板，未來會不斷的談整合、堆疊、模組化。這種尺寸與整合性的急劇變化，促使電路板密度、立體應用同步提升。

IC Package Trend

Electronic devices miniaturize, most critical point still the package & feature size shrinkage. Before we talked about lead frame transfer to substrate package. But from now on we have to focus on integration; stacking; modulization. Down size & integration ramping extremely fast, pushed PCB's density & 3D architecture grow in the same time.

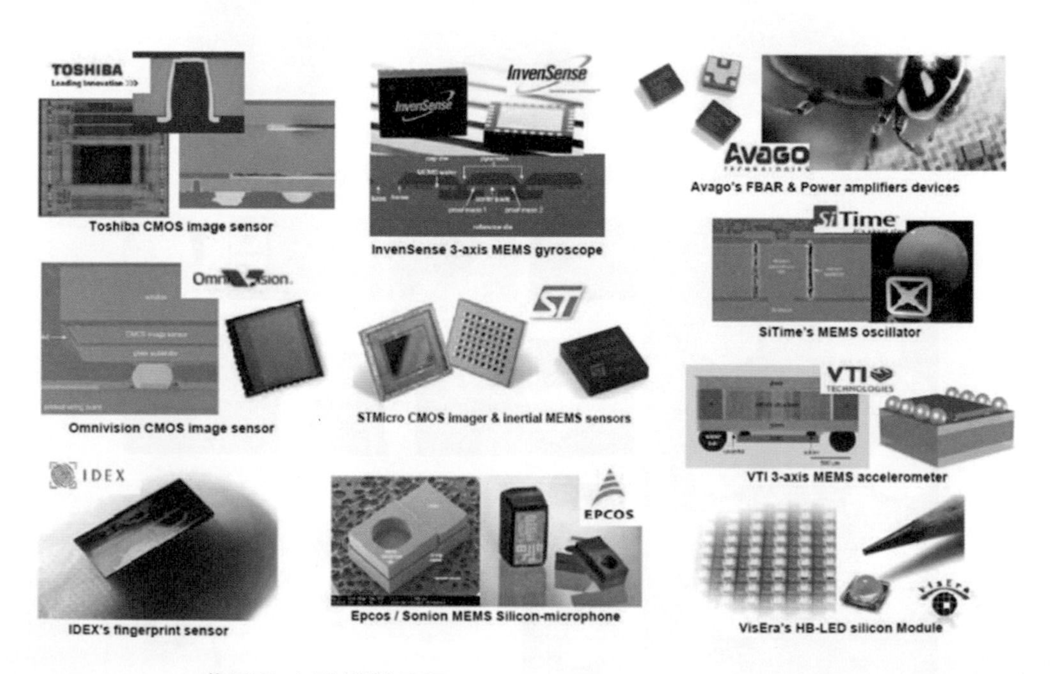

典型的立體構裝應用 / Typical 3D package application

為何需要 HDI 的技術

可攜式產品快速發展，個人化設計推陳出新，促使行動多樣化需求提高。而電路板所應配合的，當然就是更高密度、精確互連結構，這就是一般業者不斷追求的所謂 HDI 技術。

Why we need HDI process

Mobile devices progress fast, personal demand fitting makes products design getting diversification. The PCB technology have to match higher density & registration architecture, this is the never ending approach & so called HDI technology.

增加襯墊密度
Increase pad density
- More pad per area
- Merge pad and via

增加繞線密度
Increase routing density
- More layer per panel
- More line per channel

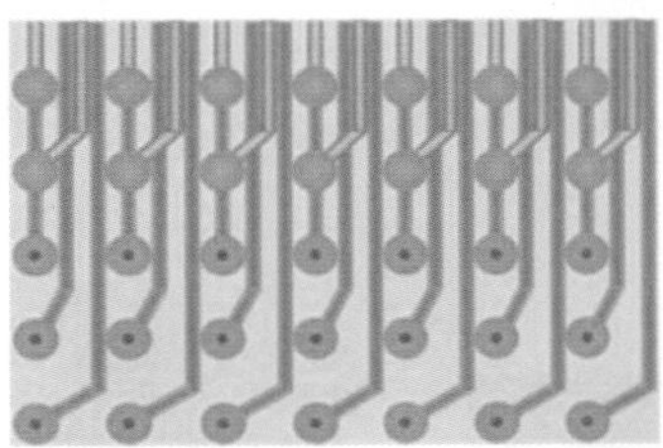

高密度互連結構
HDI structure:
Sequence Build Up
Via on PTH
Via in Pad
Stack Via

高層板結構
High Layer Count
Thin Core
Thick Board
Fine Line

相同面積填入更多元素，必然需要縮小元件尺寸增加層數

Same area more components, shrink parts and increase layer count is nature

電路板的疊合型式

電路板疊合型式在垂直方向的結構，整體發展會從傳統全通孔走向多次壓合多層孔結構，再朝向盲埋孔結構型式，發展，逐漸會讓構裝與電路板界線模糊化，走入埋入元件的結構，也就是進階的 HDI。這種趨勢，可以大幅提高繞線密度及組裝密度。

Board Structure Trend

Board design architecture will grow in Z direction. And through hole get in to sequence lamination. After micro via, embedded in components will be part of selection. Electronic package & substrate will be mixing. In some view point, this might called advance HDI technology. This trend can highly increase the routing density and assembly density.

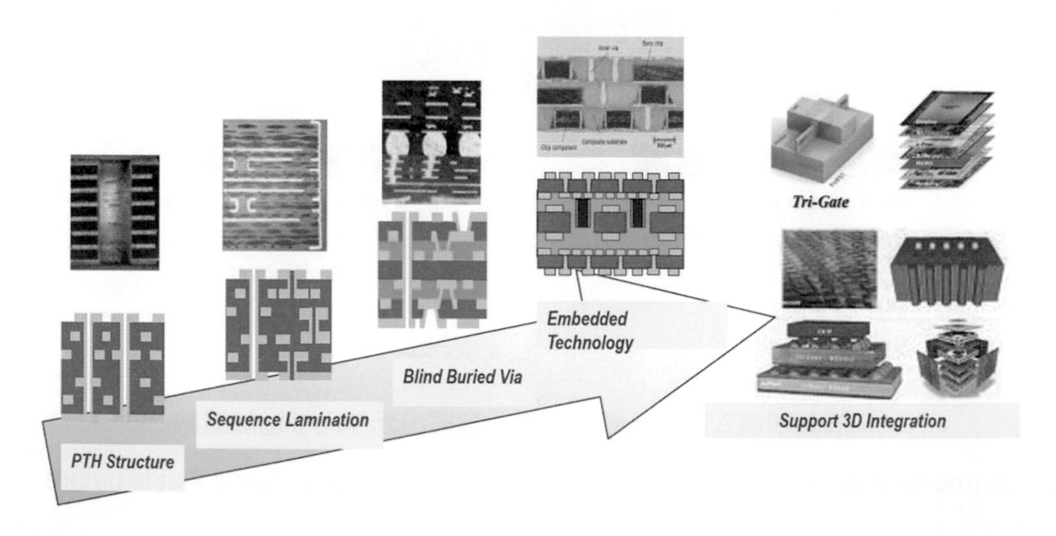

電路板的堆疊結構趨勢 / Board Stacking Trend

典型線路設計趨勢

爲了更高密度設計的發展，電路板的線路密度會逐年堆進。雖不敢說可以突破微米的水準，但是高階的構裝載板確實已經接近這個水準。過去線路邊緣粗糙的狀態，也因爲公差的減少而必須採用低稜線材料與技術。

Typical pattern design trend

For high density layout, finer pattern will ramp year by year. Nobody dare to say that fine line level will reach less than micro meter level. But some advance substrate designs are not far away from that. The conventional pattern roughing side wall won't be the case anymore. For less tolerance demanding, maker have to introduce low profile material and processing technology.

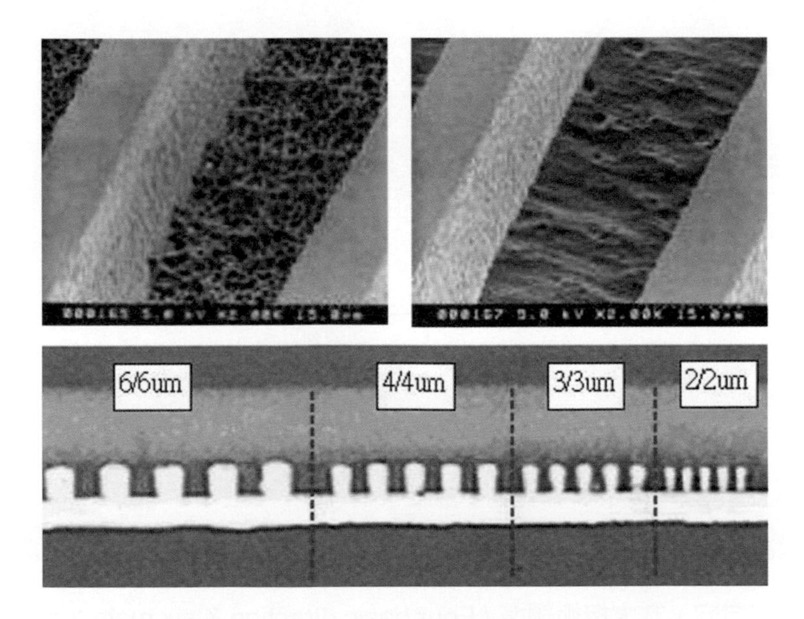

低稜線材料與細線路技術將成為未來的基本功

Low profile & fine line will be the basic technology for future

跨領域的思維

電子產業，其實並沒有明確的領域界線。過去因為產品的整合度較低，所以電路板業者可以自己發展，再進入構裝、組裝領域與系統搭配。未來的世界，任何可以對產品有貢獻的作法，都可能被導入進行系統整合。

產業界線會模糊化，分工也會隨需要而變化。筆者常說技術發展，可以用簡單的文字關係，提供一個未來技術大方向的固定思考。而產業分工的關係，也如後圖的連結，可以有彈性變化的思考。

True	Size	Fine pattern & less tolerance
	Position	Higher registration & A/R
	Selection	Better resolution & contrast
	Structure	3D architecture & MEMS

Architecture creation technology for mass production:
Material; Equipment; Tooling; Processes; Metrology; Integration

四大基本方向，六大技術領域 / Four basic direction & six main technology

SMT pad design trend

Electronic industry doesn't have clear boundary for territory identity. Early days, products integration was low, PCB maker can manufacturing themselves & then send for packaging; assembly after. In the future, anything benefit for end product will be called for joint the end products developing.

It's hard to say who has to response for what. The relationship will be complicate & flexible. As the author often said, organize some simple phrase can point out the future technology firmly direction. The following diagram can be a reference for readers do the flexible thinking.

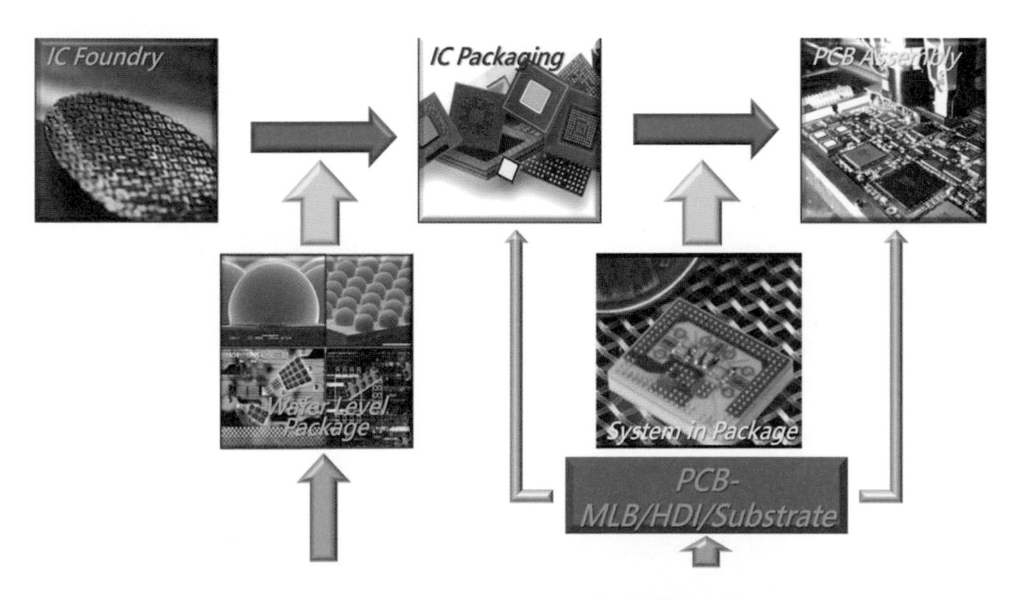

供應鏈可能的變化，相關埋入技術是立體構裝方案的一部份

Supply chain might change, Embedded is one of the 3D solution.

（請由此線剪下）

讀者回函卡

填寫日期：　／　／

姓名：＿＿＿＿ 生日：西元＿＿年＿＿月＿＿日 性別：□男 □女

電話：（ ）＿＿＿＿ 傳真：（ ）＿＿＿＿ 手機：＿＿＿＿

e-mail：（必填）＿＿＿＿

註：數字零，請用 Φ 表示，數字 1 與英文 L 請另註明並書寫端正，謝謝。

通訊處：□□□□□

學歷：□博士 □碩士 □大學 □專科 □高中・職

職業：□工程師 □教師 □學生 □軍・公 □其他

學校／公司：＿＿＿＿ 科系／部門：＿＿＿＿

・需求書類：

□ A. 電子 □ B. 電機 □ C. 計算機工程 □ D. 資訊 □ E. 機械 □ F. 汽車 □ I. 工管 □ J. 土木

□ K. 化工 □ L. 設計 □ M. 商管 □ N. 日文 □ O. 美容 □ P. 休閒 □ Q. 餐飲 □ B. 其他

・本次購買圖書為：＿＿＿＿ 書號：＿＿＿＿

・您對本書的評價：

封面設計：□非常滿意 □滿意 □尚可 □需改善，請說明＿＿＿＿

內容表達：□非常滿意 □滿意 □尚可 □需改善，請說明＿＿＿＿

版面編排：□非常滿意 □滿意 □尚可 □需改善，請說明＿＿＿＿

印刷品質：□非常滿意 □滿意 □尚可 □需改善，請說明＿＿＿＿

書籍定價：□非常滿意 □滿意 □尚可 □需改善，請說明＿＿＿＿

整體評價：請說明＿＿＿＿

・您在何處購買本書？

□書局 □網路書店 □書展 □團購 □其他＿＿＿＿

・您購買本書的原因？（可複選）

□個人需要 □幫公司採購 □親友推薦 □老師指定之課本 □其他＿＿＿＿

・您希望全華以何種方式提供出版訊息及特惠活動？

□電子報 □ DM □廣告（媒體名稱＿＿＿＿）

・您是否上過全華網路書店？（www.opentech.com.tw）

□是 □否 您的建議＿＿＿＿

・您希望全華出版那方面書籍？＿＿＿＿

・您希望全華加強那些服務？＿＿＿＿

～感謝您提供寶貴意見，全華將秉持服務的熱忱，出版更多好書，以饗讀者。

全華網路書店 http://www.opentech.com.tw　客服信箱 service@chwa.com.tw

2011.03 修訂

親愛的讀者：

感謝您對全華圖書的支持與愛護，雖然我們很慎重的處理每一本書，但恐仍有疏漏之處，若您發現本書有任何錯誤，請填寫於勘誤表內寄回，我們將於再版時修正，您的批評與指教是我們進步的原動力，謝謝！

全華圖書　敬上

勘誤表

書號		書名		作者	

頁數	行數	錯誤或不當之詞句	建議修改之詞句

我有話要說：（其它之批評與建議，如封面、編排、內容、印刷品質等・・・）